T0298640

Electronic Skin:
Sensors and Systems

RIVER PUBLISHERS SERIES IN ELECTRONIC MATERIALS AND DEVICES

Series Editors:

EDOARDO CHARBON
EPFL
Switzerland

MIKAEL ÖSTLING
KTH Stockholm
Sweden

ALBERT WANG
University of California
Riverside, USA

Indexing: all books published in this series are submitted to the Web of Science Book Citation Index (BkCI), to SCOPUS, to CrossRef and to Google Scholar for evaluation and indexing

The "River Publishers Series in Electronic Materials and Devices" is a series of comprehensive academic and professional books which focus on the theory and applications of advanced electronic materials and devices. The series focuses on topics ranging from the theory, modeling, devices, performance and reliability of electron and ion integrated circuit devices and interconnects, insulators, metals, organic materials, micro-plasmas, semiconductors, quantum-effect structures, vacuum devices, and emerging materials. Applications of devices in biomedical electronics, computation, communications, displays, MEMS, imaging, micro-actuators, nanoelectronics, optoelectronics, photovoltaics, power ICs and micro-sensors are also covered.

Books published in the series include research monographs, edited volumes, handbooks and textbooks. The books provide professionals, researchers, educators, and advanced students in the field with an invaluable insight into the latest research and developments.

Topics covered in the series include, but are by no means restricted to the following:

- Integrated circuit devices
- Interconnects
- Insulators
- Organic materials
- Semiconductors
- Quantum-effect structures
- Vacuum devices
- Biomedical electronics
- Displays and imaging
- MEMS
- Sensors and actuators
- Nanoelectronics
- Optoelectronics
- Photovoltaics
- Power ICs

For a list of other books in this series, visit www.riverpublishers.com

Electronic Skin:
Sensors and Systems

Editors

Ali Ibrahim

University of Genoa, Italy
Lebanese International University, Lebanon

Maurizio Valle

University of Genoa, Italy

River Publishers

Published, sold and distributed by:
River Publishers
Alsbjergvej 10
9260 Gistrup
Denmark

www.riverpublishers.com

ISBN: 978-87-7022-216-7 (Hardback)
    978-87-7022-215-0 (Ebook)

Contents

**9 Artificial Skin and Electrotactile Stimulation for
Advanced Tactile Feedback in Myoelectric Prostheses**   **197**

*Lucia Seminara, Matija Strbac, Youssef Amin, Maurizio Valle
and Strahinja Dosen*

Preface

Electronic skin (e-skin) has gained high relevance for many application domains such as prosthetics, robotics, biomedical instrumentation, Internet of Things (IoT) systems, and industrial automation. Many relevant achievements on the e-skin development have been documented in the last two decades. The main focus has been on the e-skin sensing material and technology with the aim of mimicking human skin features, e.g., flexibility, stretchability, and time response. Regrettably, a holistic approach at system level is not yet attempted.

The proposed book aims at filling the gap and takes into account not only sensing materials but also provides a thorough assessment of the state-of-the-art system level addressing of embedded computing and data decoding, techniques for low power embedded data processing, and communication interface.

Chapter 1 provides an introduction to the topic and highlights the relevance and impact of e-skin systems by means of three examples.

Chapter 2 introduces the human sense of touch along with the physiology of the human skin and provides a comprehensive overview of the state-of-the-art tactile sensors. Additionally, the chapter highlights the latest breakthroughs and recent advances in the e-skins and artificial tactile sensing systems for robotic and prosthetics applications. Major technical details of transduction mechanisms in the e-skins such as piezoresistivity, capacitance, and piezoelectricity are also described with their merits and demerits. Furthermore, the chapter reports the requirements, challenges, expected improvements, and future perspectives for tactile sensing technologies.

Chapter 3 discusses the major requirements and challenges of tactile sensors. The chapter reports the major achievements in the field of tactile sensor devices and technology. The chapter also highlights novel concepts to provide additional features and a full tactile feedback from sensors, with the final goal of mimicking the surprising capabilities of the human skin to sense the surrounding environment.

Chapter 4 describes the recent developments of optical-based soft tactile sensing. As a case study, a comparison between machine learning (ML) and analytical approaches – to decode tactile information in a continuum soft optical waveguide – is presented.

Chapter 5 reports the development of a novel, large area soft artificial skin with integrated fiber Bragg grating (FBG) sensors for the robotic perception of physical interaction. Specifically, by combining the FBG technology and ML techniques (fully connected neural network, FCNN), the considered artificial skin permits to simultaneously solve contact location and force, thus fostering applications in collaborative robotics.

Chapter 6 presents a survey of the existing algorithms and methods for tactile data processing. The proposed algorithms and tasks include ML, deep learning, feature extraction, and dimensionality reduction. The chapter also provides guidelines for selecting appropriate hardware platforms for the embedded implementation. Different algorithms are compared in terms of computational load and hardware implementation requirements. In addition, the chapter introduces a case study for touch modality classification.

Chapter 7 reports the state-of-the-art approximate computing techniques (ACTs) employed at the circuit level for embedded ML algorithms in the e-skin systems. The chapter demonstrates the feasibility of the ACTs in the e-skin systems by implementing approximate arithmetic circuits, mainly multipliers and adders, in the embedded ML algorithms with the aim of enhancing the overall efficiency.

Chapter 8 deals with data communication and transmission subsystem. The chapter introduces the working principles of optical communication systems by describing the coding and decoding procedures that use a combination of optical and analogue/digital electronic architectures designed to be integrated in complimentary metal–oxide semiconductor (CMOS) technology. As a case study, two applications concerning tactile sensor data communication in prosthetic systems and neural implantable devices are reported and discussed.

Chapter 9 proposes a novel concept of a high-bandwidth feedback interface that relies on advanced sensing and stimulation to convey a large amount of information to the prosthesis user. The interface comprises the e-skin covering the prosthesis with a dense network of tactile sensors and a compact stimulation device delivering electrical current pulses through a matrix electrode with multiple conductive pads. The state-of-the art for the implementation of the proposed concept are reviewed. This includes biomimetic e-skins suitable for the application in a wearable scenario, stimulation systems integrating a demultiplexing circuit to distribute electrical pulses, and flexible electrodes with arbitrary shape, size, and distribution of conductive pads. Finally, the challenges in selecting feedback variables and mapping of these variables into stimulation parameters are addressed.

Acknowledgement

The editors are very grateful to all (Ph.D. students, post-doc researchers, technicians, etc.) who collaborated over the last 15 years with the COSMIC laboratory (www.cosmic_lab.diten.unige.it) at the University of Genoa.

Many thanks to all the authors who have contributed to this book and to all those who have supported us.

Ali Ibrahim
Maurizio Valle

List of Contributors

Ali Ibrahim, *University of Genoa, Italy; Lebanese International University, Lebanon; E-mail: ali.ibrahim@edu.unige.it*

Andrea Adami, *Fondazione Bruno Kessler, Center for Materials and Microsystems, Italy; E-mail: andadami@fbk.eu*

Andrea Aliperta, *The BioRobotics Institute of Scuola Superiore Sant'Anna and Department of Excellence in Robotics and AI of Scuola Superiore Sant'Anna, Pontedera, Italy; E-mail: andrea.aliperta@santannapisa.it*

Andrea De Marcellis, *Electronic and Photonic Integrated Circuits and Systems (EPICS) Laboratory, Department of Industrial and Information Engineering and Economics, University of L'Aquila, Italy; E-mail: andre.demarcellis@univaq.it*

Calogero Maria Oddo, *The BioRobotics Institute of Scuola Superiore Sant'Anna and Department of Excellence in Robotics and AI of Scuola Superiore Sant'Anna, Italy; E-mail: calogero.oddo@santannapisa.it*

Edoardo Sinibaldi, *Department of Mechanical and Aerospace Engineering Sapienza University of Rome, Italy; E-mail: edoardo.sinibaldi@iit.it*

Eduardo Palermo, *Center for Micro-BioRobotics, Istituto Italiano di Tecnologia, Italy; E-mail: eduardo.palermo@uniroma1.it*

Egidio Falotico, *The BioRobotics Institute, Scuola Superiore Sant'Anna, Italy*

Elia Palange, *Electronic and Photonic Integrated Circuits and Systems (EPICS) Laboratory, Department of Industrial and Information Engineering and Economics, University of L'Aquila, Italy; E-mail: elia.palange@univaq.it*

Emiliano Schena, *Research Unit of Measurements and Biomedical Instrumentation, Center for Integrated Research, Universita' Campus Bio-Medico di Roma, Italy; E-mail: e.schena@unicampus.it*

Giulia Fransvea, *The BioRobotics Institute of Scuola Superiore Sant'Anna and Department of Excellence in Robotics and AI of Scuola Superiore Sant'Anna, Italy; E-mail: giulia.fransvea@santannapisa.it*

Giuseppe Terruso, *The BioRobotics Institute of Scuola Superiore Sant'Anna and Department of Excellence in Robotics and AI of Scuola Superiore Sant'Anna, Italy; E-mail: giuseppe.terruso@santannapisa.it*

Guido Di Patrizio Stanchieri, *Electronic and Photonic Integrated Circuits and Systems (EPICS) Laboratory, Department of Industrial and Information Engineering and Economics, University of L'Aquila, Italy; E-mail: guido.dipatriziostanchieri@graduate.univaq.it*

Hamoud Younes, *University of Genoa, Italy; Lebanese International University, Lebanon; E-mail: hamoud.younes@edu.unige.it*

Hoda Fares, *University of Genoa, Italy; E-mail: hoda.fares@edu.unige.it; hodafares1@gmail.com*

Jessica D'Abbraccio, *The BioRobotics Institute of Scuola Superiore Sant'Anna and Department of Excellence in Robotics and AI of Scuola Superiore Sant'Anna, Italy; E-mail: jessica.dabbraccio@santannapisa.it*

Leandro Lorenzelli, *Fondazione Bruno Kessler, Center for Materials and Microsystems, Italy*

Luca Massari, *The BioRobotics Institute of Scuola Superiore Sant'Anna and Department of Excellence in Robotics and AI of Scuola Superiore Sant'Anna, Italy; E-mail: luca.massari@santannapisa.it*

Lucia Beccai, *Center for Micro-BioRobotics, Istituto Italiano di Tecnologia, Italy; E-mail: lucia.beccai@iit.it*

Lucia Seminara, *CosmicLab, Department of Electrical, Electronic, Telecommunications Engineering and Naval Architecture (DITEN), University of Genoa, Italy; E-mail: lucia.seminara@unige.it*

Marco Faccio, *Electronic and Photonic Integrated Circuits and Systems (EPICS) Laboratory, Department of Industrial and Information Engineering and Economics, University of L'Aquila, Italy; E-mail: marco.faccio@univaq.it*

Mario Osta, *DITEN-COSMIC Lab, University of Genova-Italy, Italy; E-mail: mario.osta@edu.unige.it*

Martina Zaltieri, *Research Unit of Measurements and Biomedical Instrumentation, Center for Integrated Research, Universita' Campus Bio-Medico di Roma, Italy; E-mail: m.zaltieri@unicampus.it*

Massimo Totaro, *Center for Micro-BioRobotics, Istituto Italiano di Tecnologia, Italy*

Matija Strbac, *Tecnalia Serbia Ltd., Serbia; E-mail: matija.strbac@tecnalia.com*

Matteo Lo Preti, *Center for Micro-BioRobotics, Istituto Italiano di Tecnologia, Italy; The BioRobotics Institute, Scuola Superiore Sant'Anna, Italy; E-mail: matteo.lopreti@iit.it*

Maurizio Valle, *CosmicLab, Department of Electrical, Electronic, Telecommunications Engineering and Naval Architecture (DITEN), University of Genoa, Italy; DITEN-COSMIC Lab, University of Genova-Italy, Italy; E-mail: maurizio.valle@unige.it*

Mohamad Alameh, *University of Genoa, Italy; E-mail: mohammad.alameh@edu.unige.it*

Mostafa Rizk, *Lebanese International University, Lebanon; E-mail: mostafa.rizk@liu.edu.lb*

Strahinja Dosen, *Department of Health Science and Technology, The Faculty of Medicine, Aalborg University, Denmark; E-mail: sdosen@hst.aau.dk*

Tamás Czimmermann, *The BioRobotics Institute of Scuola Superiore Sant'Anna and Department of Excellence in Robotics and AI of Scuola Superiore Sant'Anna, Italy; E-mail: tamas.czimmermann@santannapisa.it*

Youssef Amin, *CosmicLab, Department of Electrical, Electronic, Telecommunications Engineering and Naval Architecture (DITEN), University of Genoa, Italy; E-mail: youssef.amin@edu.unige.it*

List of Figures

xxi

List of Tables

List of Abbreviations

ACME	Asymmetric compute-memory extension
ACT	Approximate computing technique
ADC	Analog-to-digital converter
ANFIS	Adaptive neuro-Fuzzy inference system
Approx-BW	Approximate Baugh-Wooley
ASIC	Application-specific integrated circuit
ASM	Alphabet set multiplier
AXA	Approximate XNOR-based adder
BAM	Broken-array multiplier
BER	Bit error rate
BLMS	Block least mean
BRAM	Block random access memory
CART	Classification and regression trees
CCD	Charge-coupled device
CMOS	Complementary metal–oxide semiconductor
CNN	Convolutional neural network
CNS	Central nervous system
CORDIC	Coordinate rotational digital circuits
CPU	Central processing unit
CSD	Canonical signed digit
CSE	Common subexpression elimination
DCNN	Deep convolutional neural network
DCT	Discrete cosine transform
DEFPI	Diaphragm-based extrinsic Fabry-Perot interferometer
DL	Deep learning
dLAC	Deep-learning accelerators
DNN	Deep neural network
DSP	Digital signal processor
DT	Decision tree
ECU	Electronic computing unit
EFPI	Extrinsic Fabry-Perot interferometer

EIT	Electrical impedance tomography
EPU	Embedded processing unit
E-skin	Electronic skin
ETA	Error tolerant adder
FBG	Fibre Bragg grating
FF	Flip-flop
FFT	Fast Fourier transform
FIR	Finite impulse response
FIS	Fuzzy inference system
FPGA	Field programmable gate array
FSM	Finite state machine
FTIR	Frustrated total internal reflection
GPU	Graphics processing unit
ICA	Independent component analysis
IPP	Input pre-Processing
KNN	K-Nearest neighbour
Kulk	Kulkarni
LCP	Linearly compressed pages
LDA	Linear discriminant analysis
LEDs	Light-emitting diodes
LOA	Lower-part-OR Adder
LSB	Least significant bits
LUT	Lookup table
MCM	Multiple constant multipliers
MCU	Microcontroller unit
MEM	Microelectromechanics
ML	Machine learning
MSB	Most significant bits
MUX	Multiplexer
NB	Naïve Bayes
OFETs	Organic field-effect transistors
OPVs	Organic photovoltaics
PCA	Principal component analysis
PCBs	Printed circuit boards
PD	Photodiode
PDMS	Polydimethylsiloxane
PLL	Phase-locked loop
PNS	Peripheral nervous system
POF	Plastic optical fibres

PULP	Parallel ultralow-power platform
PVDF	Polyvinylidene fluoride or polyvinylidene difluoride
PYNQ	Python productivity for Zynq
RE	Relative error
ROBA	Approximate multiplier based on rounding technique
Si	Silicon
SNR	Signal-to-noise ratio
STAxCache	Spintronic approximate cache
SV	Support vector
SVD	Singular value decomposition
SVM	Support vector machine
TIA	Transimpedance amplifier
TIR	Total internal reflection
TPU	Tensor processing unit
VCSEL	Vertical cavity surface emitting laser
VHDL	Verilog hardware descriptive language

1

Electronic Skin Systems

Ali Ibrahim[1,2,*] **and Maurizio Valle**[1]

[1]University of Genoa, Genoa, Italy
[2]Lebanese International University, Lebanon
E-mail: ali.ibrahim@edu.unige.it; maurizio.valle@unige.it
*Corresponding Author

The skin is one of the main organs of the human body and it is responsible for many important functions such as sensing the cutaneous stimuli and protecting the inner body organs. The development of electronic skin (e-skin) systems involves many different research areas; the application domains are numerous and relevant, e.g., robotics, teleoperated systems, biomedical systems, virtual and augmented reality, autonomous systems, etc. This chapter briefly introduces e-skin systems, highlighting three relevant use cases. First, the chapter starts with a definition of the e-skin system highlighting its importance in different domains. Then, it describes how the e-skin system is integrated into humanoid robotics namely in the iCub robot [1]. After that, we present how the use of e-skin systems may enable new technologies reporting the case of its use in telemanipulation [2]. Finally, an example for the use of the e-skin system to restore the sense of touch in prosthetic devices [3] is provided. Through these three concrete examples, the chapter aims to present the motivations, goals, and advantages of e-skin systems.

1.1 Introduction

Inspired by the structure of human skin, the electronic skin (e-skin) and its application in many and diverse application domains has attracted many researchers. The implementation of an e-skin system has been targeted either to mimic the sensing functions of human skin or to address specific applications such as robotics [4], health monitoring [5], human machine

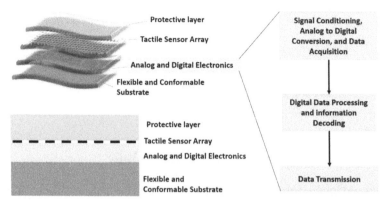

Figure 1.1   Electronic skin system structure (left) and functions (right). Adapted from [9].

interfaces [6], prosthetics [7], etc. E-skin is used to extract contact information related, e.g., to grasping, slipping, or sticking and also to tune the contact force based on texture and hardness/softness of objects. E-skin feedback can provide the sensory feeling to prosthetic users and amputees with the goal of restoring essential functions such as grip and slip and touch and pain. It can detect the pressure from the surrounding object exerted on a robot. If integrated with multifunctional sensors, e-skin can also be used to monitor essential biometric parameters, e.g., temperature, blood pressure, sugar level, etc. [8].

The e-skin system can be defined as a hybrid arrangement composed of different layers. A general illustration of the structure (left) and of the functions (right) of the e-skin system is shown in Figure 1.1 (adopted from [9]). The protective layer, usually made of polymers (e.g., PDMS), protects the sensing layer from damages when contacting the environment. The electronic layer hosts analog and digital circuits: it should conform to the usually curved surface of the substrate layer which could be rigid or soft.

The main functional components of the e-skin system are shown in Figure 1.1, right: (1) sensing materials and tactile sensor array, (2) front end electronics for signal conditioning, analog to digital conversion, and data acquisition, (3) embedded computing and data decoding unit implementing tactile data processing, and (4) communication interface in charge for communicating the touch information to the further level of the overall system.

E-skin systems operate in a seamless way, are autonomous from computational and energetic point of view, and take decisions implementing sophisticated and complex data processing algorithms (e.g., machine Learning).

This chapter aims to exemplify e-skin systems with three relevant use cases, namely humanoid robotics, telemanipulation, and prosthetics. Through three concrete examples, the chapter will present the motivations, aims, and advantages of e-skin systems. Section 1.2 will show how a humanoid robot (i.e., iCub) can benefit from whole body e-skin. E-skin system in telemanipulation as proposed in the EU2020 TACTILITY project is described in Section 1.3. In Section 1.4, e-skins for upper limb prosthetics are reported and some examples about the restoration of tactile information corresponding to some essential functions are described. Finally, conclusions are drawn in Section 1.5.

1.2 Integration of E-skin in iCub Robot

E-skin in robotics is essential for performing three principle tasks: (1) manipulation (perception for action), (2) exploration (action for perception), and (3) reaction/haptics (action–reaction). Robots are expected to have such abilities and adapt to the environment [10].

The e-skin system of iCub [1] employs capacitive distributed pressure sensors consisting of two electrodes with a soft dielectric in between. The capacitance at the output of the sensor changes as function of the distance between the electrodes. The e-skin system is constituted of interconnected triangular modules (i.e., flexible printed circuit boards (PCBs) with the electronics below) forming a mesh of sensors that are able to cover non-flat surfaces. It consists of the PCBs having triangular shape, hosting 12 sensors and a capacitance to digital converter AD7147 from analog devices. It obtains 12 capacitance measurements and sends them through a serial bus. Four wires for the serial bus are connected to the PCB. Curved surfaces can be covered due to the flexibility of triangles as well as to the connections among them. The next subsections provide some practical examples on the integration of the e-skin into the iCub robot, especially on its fingertips, palm, and forearm.

1.2.1 E-skin on iCub Fingertips

A dedicated e-skin system that resembles a human fingertip has been proposed in [1]. Each fingertip has a round shape with the size of 14.5 mm long and 13 mm wide. Figure 1.2 shows the flexible PCB having the ability to be wrapped, making the fingertips. They are connected to small boards on the

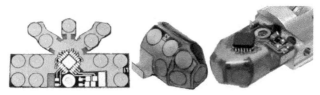

Figure 1.2   E-skin on iCub fingertips. From [1]. Reprinted with permission from IEEE.

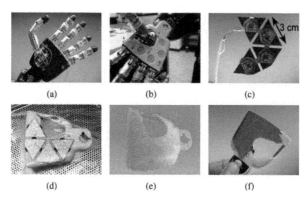

Figure 1.3   E-skin on iCub Palm. From [1]. Reprinted with permission from IEEE.

back of the hand. These boards relay the data from all the fingertips to one microcontroller board located in the iCub forearm.

1.2.2 E-skin on iCub Palm

The implementation steps of the e-skin on the iCub palm are reported in Figure 1.3. The iCub palm is made from carbon fiber with a support cover for the sensor. The PCB includes a capacitance to digital converter chip.

1.2.3 E-skin on the iCub Forearm

The same process as for the palm has been followed, and it is illustrated in Figure 1.4. Figure 1.5 presents the iCub arm covered with the e-skin system. Each arm is composed of 384 sensing points distributed as follows: PCBs × sensing elements = 23 × 12 in the forearm, 4 × 12 in the palm, and 5 × 12 in the fingertips.

iCub has been taught in [11] to perform the grasp by providing an intuitive notion of force in addition to the implicit knowledge of the kinematics necessary for adaptation. iCub has successfully achieved a successful grasp adaptation when the contact changes for multiple objects [11].

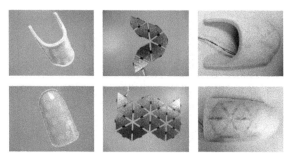

Figure 1.4 E-skin on the iCub Forearm. From [1]. Reprinted with permission from IEEE.

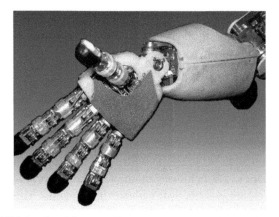

Figure 1.5 iCub hand and forearm. From [1]. Reprinted with permission from IEEE.

In [12], a human teacher has taught iCub the shape of an object by guiding its hand along the object especially on the part to be grasped (e.g., the handle of a jar). The e-skin has allowed iCub to explore an object using its two hands with above 90% success rate of identification [12]. iCub is able to interact with the environment and perceive the objects [13, 14]. For instance, he is able to follow the contour of touched linear and circular objects [13]. The e-skin system has permitted iCub to interact with objects with different shapes, to afford packs and lift them, to achieve an impedance control while holding an object, to apply actions, lifting, and swinging [15]. To provide an example, the authors in [1] presented an experiment demonstrating the effectiveness of the tactile feedback in improving the grasp of a fragile plastic cup for iCub. A plastic cup is placed in the iCub hand and the grasp has been started following two scenarios: with and without tactile feedback. Figure 1.6(left) shows how iCub has crushed the cup when no tactile feedback is provided,

Figure 1.6   iCub is grasping a fragile plastic cup. (Left) Without tactile feedback, iCub crushes the cup. (Right) With tactile feedback, it grasps the cup without deforming it. From [1]. Reprinted with permission from IEEE.

whereas, with the help of tactile feedback in Figure 1.6(right), he was able to grasp the cup gently without deforming it.

1.3 E-skin in Telemanipulation: of the EU2020 TACTILITY Project

The bilateral teleoperation systems illustrated in Figure 1.7 usually rely on console like controller or joystick with force or haptic feedback. The feeling of embodiment is not achieved by the feedback mechanism and only provide intuitive navigation scheme for planar movements. This issue could be addressed by implementing the telemanipulation control using the TACTILITY gloves. The gloves may provide accurate tracking for the kinematic movement allowing to control precisely the industrial robots by performing hand gestures and by employing the tactile feedback feature. This feature is based on gathering the tactile information through the e-skin system and delivering them in real time to the teleoperator after being processed onto electrotactile feedback.

Integrating the e-skin system and providing the tactile feedback in telemanipulation systems may enable the teleoperator to be immersed into the remote environment controlling the industrial robot as an extension of their body. This will have its impact on increasing the productivity and the

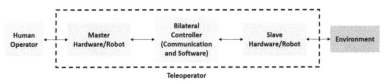

Figure 1.7   Block diagram for the bilateral teleoperation system.

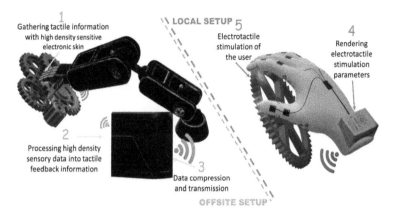

1
Gathering tactile information
with high density sensitive
electronic skin

LOCAL SETUP 5
Electrotactile
stimulation of
the user

4
Rendering
electrotactile
stimulation
parameters

2
Processing high density
sensory data into tactile
feedback information

3
Data compression
and transmission

OFFSITE SETUP

Figure 1.8 Telemanipulation application scenario: TACTILITY project concept [2].

effectiveness of the teleoperator while providing safety and satisfaction to him in their workplace.

The second application scenario, namely telemanipulation, is taken from the EU 2020 TACTILITY project [2]. The concept of the system is depicted in Figure 1.8: (1) a high-density e-skin system to gather tactile information; (2) the local processing unit for data acquisition and touch information extraction from raw sensed data; (3) the extracted information is compressed to a form suitable for communication to the remote user. Finally, the electrotactile stimulation parameters are rendered (4) and delivered to the remote user (5). The user interacts with an environment which is not fully controlled and known as in the case of artificially generated virtual reality. Instead, the user sees and touches the records from cameras and e-skin which increases the complexity of this approach.

1.4 E-skin in Upper Limb Prostheses

Recent technological advances can aid the prosthetic user ability to regain the lost functions by restoring the perception and the meaningful tactile information [16]. As a third application scenario, we consider here the case of integrating the e-skin system into the upper limb prostheses to enable the restoration of sense of touch. This could provide improvements in assisting the prosthetic user in its daily life and improves the prosthesis embodiment. The system substitutes the lost tactile sensation with the artificial e-skin and provides to the patient the essential information such as grip and slip, touch and pain, or object recognition and texture discrimination.

The regulation of grip and the slip prevention are considered as subliminal reflexes that can be restored in prosthesis. To this aim, the authors in [3] have developed a feedback system for upper limb prosthetic using piezoresistive tactile sensors. The system is able to attenuate the strength of grip and detect the slip events by measuring the contact force resulting in minimizing the number of fallen and broken objects.

Various applications could be implemented by the e-skin system to extract meaningful information such as object size and shape discrimination [17], roughness classification [18], curvature detection [19], or texture discrimination [20, 21]. For instance, the Izhikevich neuron model has been used in [22] to stimulate the median nerve with a spike train through transcutaneous electrical nerve stimulation. This approach has successfully enabled texture discrimination from various coarsenesses. In the following subsections, two approaches for the integration of e-skin into prosthetics [3, 23] are reported (see Figure 1.9).

1.4.1 Piezoelectric-based E-Skin

Figure 1.9(a) introduces the development of e-skin system in the sensory feedback loop for electro-stimulation case. An array of 4×4 tactile sensors has been used for fingertips based on screen printed piezoelectric polymer (PVDF) with sensor diameter of 1 mm and pitch of 1.6 mm. Whereas, a Thru mode commercial FSR sensor array (MS9723) consisting of 16×10 piezoresistive sensors has been employed for the palm (8×5 cm^2). The system includes an interface electronics for data acquisition and signal pre-processing. A fully programmable wireless multichannel electro-stimulator with flexible electrodes is used to convey the decoded information to the prosthetic user.

The main goal was to perceive the stimulation corresponding to the applied touch on the e-skin system. Right lines and some complex shapes were applied; e.g., geometries and letters. Users should identify the touch on the e-skin surface by only receiving information through electro-stimulation. This helps in understanding the capability of the human brain to interpret the sensory feedback through the e-skin stimulation system.

1.4.2 Piezoresistive-based E-Skin

Figure 1.9(b) illustrates the work presented in [3]. The system transforms the tactile sensor readings into neuromorphic spikes to enable the perception of touch and pain through nerve stimulation. Stimulation parameters have

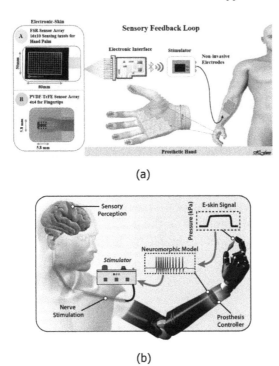

(a)

(b)

Figure 1.9 E-skin integration in prosthetic feedback systems. (a) non-invasive approach for restoring the sense of touch in prosthetics by employing an e-skin on palm and fingertips. The e-skin system is composed of the sensor matrix (on the left), the interface electronics (in the middle) sending the commands to the stimulator, and the stimulator in charge of generating electrical pulses to be delivered to the user. From [23]. Reprinted with permission from IEEE. (b) A sensory neuroprosthesis transforming into neuromorphic spikes the e-skin readings. When interacting with a sharp object, the amputee feels pain and the object is released accordingly. From [24]. Reprinted with permission from AAAS.

been discovered and quantified eliciting noxious (painful) and innocuous (non-painful) tactile perceptions through transcutaneous electrical nerve stimulation. Piezoresistive and conductive fabrics were used to design the sensor to measure the pressure applied on the e-skin surface. A rubber layer of 1 mm size is located between the top (epidermal) and bottom (dermal) layers to distribute the loads at the moment of grasp. Each fingertip includes three tactile taxels with 1.5 cm^2 of combined sensing area. Layering the sensor implies a variation in the e-dermis output when loads are applied. This can be exploited to extract more useful information such as object curvature and pressure distribution.

1.5 Conclusion

This chapter has presented three important application scenarios where the e-skin system has been integrated namely humanoid robotics, telemanipulation, and prosthetics. The e-skin integration has enabled interesting tasks that were not possible using other sensing modalities, i.e., vision and auditory. Employing the e-skin system, these applications were able to apply gentle grasping (robotics), to restore fundamental activities, i.e., touch and pain (prosthetics), and to make precise control of industrial robots (telemanipulation). Not limited to these domains, e-skin can be applied in many other interesting applications, e.g., human machine interface [25], minimally invasive surgery [26], and cyber–physical systems. To achieve that, this chapter has presented a general definition of the e-skin structure and function that covers the main essential aspect to mimic the human skin and also to respond to the applications demands.

References

[1] P. Schmitz, M. Maiolino, L. Maggiali, G. Natale, G. Cannata and G. Metta, "Methods and Technologies for the Implementation of Large-Scale Robot Tactile Sensors," in IEEE Transactions on Robotics, vol. 27, no. 3, pp. 389–400, June 2011, doi: 10.1109/TRO.2011.2132930.

[2] Available online: https://tactility-h2020.eu/access02/09/2020.

[3] L. Osborn, R. R. Kaliki, A. B. Soares, and N. V. Thakor, "Neuromimetic event-based detection for closed-loop tactile feedback control of upper limb prostheses," IEEE Trans. Haptics, vol. 9, no. 2, pp. 196–206, Apr./Jun. 2016.

[4] S. Gong, D. Lai, B. Su, K. Si, Z. Ma, L. Yap, P. Guo and W. Cheng, Highly stretchy black gold e-skin nanopatches as highly sensitive wearable biomedical sensors. Adv. Electron. Mater., 2015, 1, 1400063.

[5] C. Núñez, W. Navaraj, E. Polat and R. Dahiya, Energy-autonomous, flexible, and transparent tactile skin. Adv. Funct. Mater., 2017, 27, 1606287.

[6] H. Chen, L. Miao, Z. Su, Y. Song, M. Han, X. Chen, X. Cheng, D. Chen and H. Zhang, Fingertip-inspired electronic skin based on triboelectric sliding sensing and porous piezoresistive pressure detection. Nano Energy, 2017, 40, 65–72.

[7] A. S. Dolbashid, M. S. Mokhtar, F. Muhamad and F. Ibrahim, Potential applications of human artificial skin and electronic skin (e-skin): A review. Bioinspir. Biomim. Nanobiomater., 2018, 1, 53–64.

[8] D. Rodrigues, A. I. Barbosa, R. Rebelo, I. K. Kwon, R.L. Reis and V.M. Correlo, Skin-Integrated Wearable Systems and Implantable Biosensors: A Comprehensive Review. Biosensors 2020, 10, 79.

[9] A. Ibrahim, L. Pinna, L. Seminara, M. Valle, Achievements and Open Issues Toward Embedding Tactile Sensing and Interpretation into Electronic Skin Systems. In Material-Integrated Intelligent Systems-Technology and Applications; John Wiley & Sons, Ltd.: WeinHeim, Germany, 1 December 2017; Chapter 23; pp. 571–594.

[10] B. Shih, D. Shah, J. Li, T. Thuruthel, Y. Park, F. Iida, Z. Bao, et al. (2020). Electronic skins and machine learning for intelligent soft robots. Science Robotics, 5.

[11] Eric L. Sauser, Brenna D. Argall, Giorgio Metta, Aude G. Billard, "Iterative learning of grasp adaptation through human corrections," Robotics and Autonomous Systems, Volume 60, Issue 1, 2012, Pages 55–71, ISSN 0921-8890, doi: 10.1016/j.robot.2011.08.012.

[12] N. Sommer, M. Li and A. Billard, "Bimanual compliant tactile exploration for grasping unknown objects," 2014 IEEE International Conference on Robotics and Automation (ICRA), Hong Kong, 2014, pp. 6400–6407, doi: 10.1109/ICRA.2014.6907804.

[13] Martinez-Hernandez, U. (2014). Autonomous active exploration for tactile sensing in robotics. PhD thesis, The University of Sheffield, Sheffield, UK.

[14] N. F. Lepora, U. Martinez-Hernandez and T. J. Prescott, (2013). A solid case for active bayesian perception in robot touch. In: Biomimetic and Biohybrid Systems (pp. 154–166). Springer.

[15] Available online: https://www.iit.it/it/research/facilities/icub-tech. Access on 07/09/2020.

[16] M. M. Iskarous and N. V. Thakor, "E-Skins: Biomimetic Sensing and Encoding for Upper Limb Prostheses," in Proceedings of the IEEE, vol. 107, no. 10, pp. 2052–2064, Oct. 2019, doi: 10.1109/JPROC.2019.2939369.

[17] S. Datta, A. Khasnobish, A. Konar, D. N. Tibarewala, and R. Janarthanan, "Object shape and size recognition from tactile images," in Proc. Int. Conf. Control Commun. Comput. (ICCC), Dec. 2013, pp. 16–21.

[18] Z. Yi and Y. Zhang, "Recognizing tactile surface roughness with a biomimetic fingertip: A soft neuromorphic approach," Neurocomputing, vol. 244, pp. 102–111, Jun. 2017.

[19] W. Lee, J. Cabibihan, and N. V. Thakor, "Bio-mimetic strategies for tactile sensing," in Proc. Sensors, Nov. 2013, pp. 1–4.

[20] U. B. Rongala, A. Mazzoni, and C. M. Oddo, "Neuromorphic artificial touch for categorization of naturalistic textures," IEEE Trans. Neural Netw. Learn. Syst., vol. 28, no. 4, pp. 819–829, Apr. 2015.

[21] M. Rasouli, Y. Chen, A. Basu, S. L. Kukreja, and N. V. Thakor, "An extreme learning machine-based neuromorphic tactile sensing system for texture recognition," IEEE Trans. Biomed. Circuits Syst., vol. 12, no. 2, pp. 313–325, Apr. 2018.

[22] C. M. Oddo et al., "Intraneural stimulation elicits discrimination of textural features by artificial fingertip in intact and amputee humans," Elife, vol. 5, Mar. 2016, Art. no. e09148.

[23] M. Alameh et al., "Live Demonstration: System based on Electronic Skin and Cutaneous Electrostimulation for Sensory Feedback in Prosthetics," 2018 IEEE Biomedical Circuits and Systems Conference (BioCAS), Cleveland, OH, 2018, pp. 1–1, doi: 10.1109/BIOCAS.2018.8584710.

[24] L. E. Osborn et al., "Prosthesis with neuromorphic multilayered e-dermis perceives touch and pain," Sci. Robot., vol. 3, no. 19, 2018, Art. no. eaat3818.

[25] X. Pu, H. Guo, J. Chen et al., "Eye motion triggered self-powered mechnosensational communication system using triboelectric nanogenerator," Science Advances, vol. 3, no. 7, Article ID e1700694, 2017.

[26] S. Roy and C. Evans, "Overview of robotic colorectal surgery: current and future practical developments," World Journal of Gastrointestinal Surgery, vol. 8, no. 2, p. 143, 2016.

2

Artificial Tactile Sensing and Electronic-Skin Technologies

Hoda Fares* and Maurizio Valle

University of Genoa, Genoa, Italy
E-mail: hoda.fares@edu.unige.it; hodafares1@gmail.com;
maurizio.valle@unige.it
*Corresponding Author

Mimicking some of the remarkable characteristics of the human skin, including high sensitivity, mechanical flexibility, stretchability, and biodegradability has been one of the highly innovative research topics in both academia and industry for at least the last 30 years. Recent advances have fostered the development of artificial tactile sensing systems and opened up new opportunities for building electronic skin (e-skin) in applications such as robotics, prosthetics, and human–machine interfaces. Current tactile sensing systems have demonstrated high performance, low-cost, and ease of fabrication. However, profound issues remain unresolved and require strong interdisciplinary efforts to tackle them. This chapter primarily focuses on the strategies and technologies for the exploitation of e-skins in reconstructing tactile information. The chapter introduces the human sense of touch along with the physiology of the human skin and provides a comprehensive overview of the state-of-the-art tactile sensors. Additionally, the chapter highlights the latest breakthroughs and recent advances in the e-skins and artificial tactile sensing systems for robotic and prosthetics applications. Major technical details of transduction mechanisms in the e-skins such as piezoresistivity, capacitance, and piezoelectricity are described with their merits and demerits. Finally, requirements, challenges, improvements, and future perspectives that tactile sensing technologies need to address are reported.

2.1 Introduction

By means of the sense of touch, we learn how to interact, perceive, grasp, and manipulate objects. Over the last two decades, the increasing demand for replicating the human tactile sensing system has inspired the exploration of various tactile sensing transduction mechanisms [1, 2] and their prospective application in diverse domains, such as human–machine interfaces, prosthetics [3, 4], robotics [5, 6], virtual reality systems [7, 8], remote operation [9, 10], and touch screens [8].

Tactile sensors range from simple sensors that measure contact location to more complex ones that measure surface properties such as temperature, vibration, roughness, texture, stiffness and shape, etc. [2, 8, 9].

The human skin relies on a dense network of receptors such as mechanoreceptors, thermoreceptors, and nociceptors, to acquire and deliver spatially distributed contact information [3]. Imitating the distributed nature of the human skin poses challenges for technology; yet, it offers interesting opportunities to develop adequate artificial sensing systems. For that matter, electronic skin (e-skin) is an artificial skin, which hosts a hybrid stack of smartly arranged tactile sensing elements, interface electronics and embedded electronic systems, and a communication interface [11].

In this respect, this chapter presents the basics for tactile sensors and e-skin system concept and its ability to reconstruct tactile sensations. The human sense of touch is introduced along with the physiology of the human skin in Sections 2.2 and 2.3. Subsequently in Section 2.4, a brief review about the evolution of the tactile sensing and artificial skins is pinpointed. Finally, in Section 2.5, the e-skin system is introduced along with various transduction mechanisms regarding two application scenarios. Furthermore, a conclusion briefly states the requirements, challenges, and future developments of the e-skin.

2.2 SENSE of Touch

A huge network of nerve endings and touch receptors within the skin known as somatosensory system controls the sense of touch. This system is responsible for tactile sensations, for example feeling cold/hot, smooth/rough, pressure, tickle, itch, pain, vibration, etc. The sense of touch comprises three main subsystems, i.e. cutaneous, kinesthetic, and haptic [12, 13].

- The *cutaneous system* receives sensory inputs from the receptors embedded in the skin. As a matter of fact, the cutaneous system involves physical contact with the stimuli and provides awareness of the stimulation of the outer surface of body by means of receptors in the skin and associated somatosensory area of the central nervous system (CNS) [12, 13].

- The *kinesthetic system* receives sensory inputs from the receptors within muscles, tendons, and joints [14]. It provides information about the static and dynamic body postures (e.g. position of head, limbs, and end effectors) based on afferent information originating from the muscles, joints, and skin; and efferent copy, which correlates the muscle efferent available to the higher brain centers.
- The *haptic system* uses information about objects and events from the above-mentioned systems. It perceives heat, cold, and various stimuli that produce pain [12, 14].

Touch is undisputedly important for the development of several cognitive functions. It is bestowed with the spatio-temporal perception of external stimuli through a large number of receptors such as mechanoreceptors for pressure/vibration, thermoreceptors for temperature, and nociceptors for pain/damage that are distributed all over the body with variable density. The response to mechanical stimulus is liaised by mechanoreceptors that are embedded in the skin at different depths [12, 15].

2.3 Artificial Skin: Concept and Evolution

2.3.1 Understanding the Human Skin Physiology

The human skin acts as the protective and flexible waterproof barrier, which separates the human being from the outside environment. It relies on sensory receptors that provide information about the contact and the surrounding environment. Moreover, it is capable of sensing touch that includes mechanical stimulation, heat, and pain [13]. The skin is formed of three major layers organized from outside to inside, the epidermis, dermis, and subcutaneous fat. There are two major types of skin in humans: hairy skin and glabrous skin. The receptors found in the glabrous skin will be elaborated in this section since this skin type covers the parts of the body mainly used for tactile exploration (fingertips, palms of the hands, soles of the feet, and the lips). Compared to hairy skin, it has a thicker epidermis and a more rigid appearance. As the name suggests, it also lacks hair follicles [12, 13] .

The glabrous skin of the human hand contains 17,000 tactile units, i.e. primary afferent neurons with sensory endings in the dermis specialized

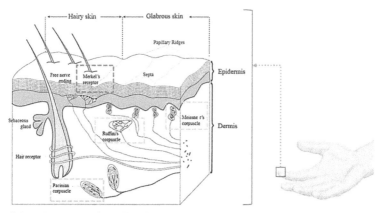

Figure 2.1   Illustration of the distribution and classification of the mechanoreceptors in the human skin (adapted from Ref. [3]).

for sensing deformations of the skin that occur when the hand interacts with objects [16]. Mechanoreceptors are sensory units distributed in the human skin to detect mechanical stimulation and to provide information about physical properties of the object and the contact between hand and object, i.e. sensations related to pressure, vibration, shape, texture, stiffness, etc. [3, 17]. Mechanoreceptors in the glabrous skin of human hands include four types: Merkel cells, Meissner corpuscles, Ruffini endings, and Pacinian corpuscles. The physical position and classification of these receptors varies across the body as shown in Figure 2.1 (adapted from Ref. [3]). Their spatial resolution is the highest at the fingertips with 1 mm [8, 18] and the lowest at the belly with 30 mm [8], as the time resolution reaches up to 700 Hz [8]. The number of receptors per square centimeter area is estimated to be 241 in the fingertips and 58 in the palm of adult humans. They are responsible for the detection of different stimulations. According to their adaptation rate, four types of mechanoreceptors are categorized into two classes: fast adapting (FA) units and slow adapting (SA) units. Further, based on their receptive fields, each class is divided into two groups: II and I. SA-I and FA-I receptors have small receptive fields, respectively 2–3 mm and 3–5 mm in diameter with a sharp border, while SA-II and FA-II receptors have large receptive fields with diffused border [16]. These four types of mechanoreceptors have different functional properties: the receptive speed, the receptive field, and the perceptive function, which are summarized in Table 2.1.

In terms of the time response, Meissner and Pacinian corpuscles are mainly responsible for rapid or dynamic stimulation, while Merkel cells

Table 2.1 Specific roles and locations of various mechanoreceptors in human skin

	Merkel Cells	Ruffini Endings	Meissner Corpuscle	Pacinian Corpuscle
Classification	SA-I	SA-II	FA-I	FA-II
Location	Shallow	Deep	Shallow	Deep
Adaptation rate	Slow	Slow	Fast	Fast
Stimuli frequency (Hz)	0.4–100	0.4–100	10–200	70–1000
Receptive field (RF)	Small and sharp (2–3 mm)	Large and diffuse is (10–15 mm)	Small and sharp (3–5 mm)	Large and diffuse ($\geq$20 mm)
Spatial resolution (mm)	0.5	7+	3–4	10+
Density (units/cm^2)	70	10	140	20
Sensory functions	Low-frequency Static forces with high resolution Pattern detection, Texture perception Tactile flow perception	Low-frequency Tension deep in the skin Stable grasp Tangential force/skin stretch	High-frequency Vibration detection Object slip, light touch, texture	High-frequency Vibrations

and Ruffini endings respond to sustained stimulation. Meissner corpuscles are sensitive to light touch, while Pacinian corpuscles tend to detect deep pressure touch and high frequency vibration. Merkel cells are sensitive to low-frequency vibration, while Ruffini endings usually respond to stretching of the skin. In terms of the location and the receptive field, Meissner corpuscles and Merkel cells concentrate in the outer layer of the skin on fingertips and have small receptive fields. On the contrary, Pacinian corpuscles and Ruffini endings are distributed more uniformly in the deeper layer of the skin (e.g. dermis) on fingers and the palm. In terms of the function of perception, Merkel cells and Pacinian corpuscles might be related to the sensation of stiffness. Merkel cells and Ruffini endings could detect slip and shape due to their response to steady pressure and skin stretch. Besides, Meissner corpuscles and Pacinian corpuscles contribute to the perception of texture, such as surface roughness, because they are sensitive to rapid vibration, which is too small to activate the other two types of mechanoreceptors [13, 15]. The spatial resolution is the smallest distance for one to distinguish two-point touch and varies across the body. It is as close as 0.5 mm on the fingertips while 7 mm on the palm [16] .

The distinct functions of various tactile units (i.e. FA-I, SA-I, FA-II, and SA-II) result in what we identify as tactile perception [19]. On the other hand, proprioception is defined as information about the configuration of the body and its position in space. Besides, from SA-II receptors, proprioception emerges from specialized receptors in the muscles (muscle spindles) and joints (Golgi tendon organs). Muscle spindles are small sensory organs distributed throughout the striated muscles of the human body, encapsulated in between the muscle fibers (called extrafusal fibers). Specialized intrafusal fibers are located in the capsule of the muscle spindles. When they are stretched, e.g. when the muscle lengthens, the muscle spindle afferent is activated. Therefore, their function is to report changes in the muscle length as well as the velocity of these changes. Golgi tendon organs are mechanoreceptors that are attached to collagen fibers in the tendons. When the muscle contracts, the collagen fibers are pulled and this activates the Golgi organ. The stronger the muscle contracts, the stronger the tension in the muscle and the tendon and the stronger the activation of the proprioceptive afferent fiber. Therefore, Golgi tendon organs report the degree of muscle tension and its change [16].

Every type of the mechanoreceptors has its own tactile-sensing mechanism in response to specific stimuli. When a tactile stimulus (i.e. mechanical stimulation) is applied to the human skin, relevant mechanoreceptors encode

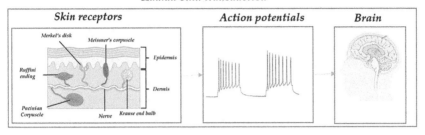

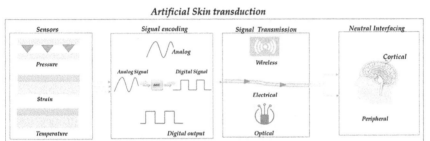

Figure 2.2 Schematic illustrates the required steps for transduction of sensory stimuli to the brain: (a) Biological skin transduction: The mechanoreceptors in the human skin encode the applied tactile stimuli into action potential, which in turn will be collected from the different presynaptic neurons and processed in the biological synapses to achieve multilevel features of tactile perception. (b) Artificial skin transduction: The various sensors detect the stimuli and translate them into signals. The collected signals are encoded and sent to the brain (adapted from Ref. [3]).

them as action potential at specific frequency. Through the nerve fibers, the generated sequence of action potentials from neurons are transmitted to the brain, which processes them into multilevel features of tactile perception [3] as shown in Figure 2.2(a). In general, artificial tactile sensors are expected to demonstrate small resolution, high sensitivity, low hysteresis, fast and linear response, wide dynamic range, and high reliability. For instance, tactile sensors should measure three-dimensional (3D) forces between 0.01 and 10 N along with a time response less than 10^{-3} s [8, 18]. A spatial resolution of 5–40 mm could be satisfactory. For high sensitivity body sites such as fingertips the spatial resolution should be around 1 mm, while for less sensitive sites it could be 5 mm especially for hand palm and shoulder. Typically, 20–60 Hz would be fine for sampling rate in daily tasks, while for a special task, such as texture recognition, a higher sampling rate of 1–2.5 kHz is necessary [15]. Robust, flexible, stretchable, and soft materials are demanded to be embedded on various 3D structures for acquiring a human-like skin. Additionally, low cost, low-power consumption, and scalability are also important for manufacture and implementation [2, 8, 18].

2.3.2 Artificial Skins

Being inspired by the unique structure of the human skin, several efforts have been made to develop artificial skins that combine a wide variety of tactile sensors to mimic the human sensory system. Artificial skins in turn would have several potential applications in human–machine interfaces robotics, artificial intelligence, prosthetics, and health monitoring technologies [8, 11]. Several types of tactile sensors are needed to match the range of signals detected by the various mechanoreceptors present in the human skin and the tactile sensing functions of these structures. The signals from the sensors require signal processing to encode them into a form that users can understand. The encoded information is sent to the brain either by direct stimulation of the CNS or PNS, using electrode arrays [3] as shown in Figure 2.2(b).

An artificial skin with sensory capabilities is commonly referred in literature as sensitive skin, smart skin, or e-skin. Such systems require integration of various sensors on a thin, flexible substrate. Usually, the e-skin is structured as a networked system of "patches" implemented as hybrid stackwise arrangements incorporating tactile sensing (i.e. mechanical into electrical transduction, signal conditioning and acquisition) and data interpretation. E-skin's structure and function varies with materials, transduction techniques, and application; e.g. e-skins scale ranges from small patches for health monitoring applications to large area patches for robotics [1, 20].

Conceivably, the earliest example of e-skin realization is from the 1970s when an artificial hand covered with skin was explored to detect grip strength, slip, and certain properties of a held object such as texture and hardness [21]. A decade later, around the 1980s, Hewlett-Packard (HP) marketed a personal computer (HP-150) that was equipped with a touchscreen, enabling users activate different functions by simply touching the display. It was the first mass-marketed electronic device exploiting the intuitive nature of human touch [22]. In 1985, General Electric (GE) built the first sensitive skin for a robotic arm using discrete infrared sensors placed on a flexible sheet at a resolution of 5 cm. The fabricated sensitive skin was proximally conscious of its surroundings, permitting the robot's arm to avoid potential obstacles and effectively move within its physical environment. Despite the robotic arm's lack of fingers and low resolution, it was capable of demonstrating the fact that electronics integrated into a membrane could allow for natural human–machine interaction [23]. In the 1990s, scientists began using flexible electronic materials to create large area, low-cost, and printable sensor sheets. The first flexible sensor sheet(s) for tactile shear force sensing was proposed by Jiang et al. by creating silicon (Si) microelectromechanical (MEM) islands by engraving thin Si wafers and integrating them on flexible polyimide foils. Around the same time, flexible arrays fabricated from organic semi-conductors began to emerge that rivaled the performance of amorphous Si [24]. Earlier, before the millennium turn, the first Sensitive Skin Workshop was held in Washington, DC under the sponsorship of the National Science Foundation and the Defense Advanced Research Projects Agency, bringing together approximately 60 researchers from different sectors of academia, industry, and government. It was revealed that there was significant industrial interest in e-skins for numerous applications, ranging from robotics to health care.

Substantial progress in the development and advancement of the e-skin has been made in recent years, and specific emphasis has been on mimicking the mechanically compliant yet highly sensitive properties of the human skin. Lacour et al. [25] developed stretchable electrodes. Kim et al. [26] altered a typically brittle material, Si, into flexible, high-performance electronics by using ultrathin (100 nm) films connected by stretchable interconnects. Someya et al. [27] fabricated flexible pentacene-based organic field-effect transistors (OFETs) for large-area integrated pressure-sensitive sheets with active matrix readout, whereas Metzger et al. [28] investigated novel pressure sensing methods using foam dielectrics and ferro-electrets [29] integrated with FETs. Bao's group studied the use of microstructured elastomeric dielectrics for highly sensitive capacitive pressure sensors [30] and developed a composite conductive elastomer revealing repeatable self-healing and mechanical force sensing capabilities [31]. Other groups developed stretchable optoelectronics, including light-emitting diodes (LEDs) [32, 34] and organic photovoltaics (OPVs) [33, 35] for integration within the e-skin.

Kim et al. [34] demonstrated a stretchable artificial skin based on ultra-thin single crystalline silicon nanoribbons that integrate strain, pressure, and temperature sensor arrays as well as associated humidity sensors, electrore-sistive heaters, and stretchable multielectrode arrays for nerve stimulation. This collection of stretchable sensors and actuators facilitate highly localized mechanical and thermal native-skin-like perception in response to external stimuli, thus providing unique opportunities for emerging classes of prostheses and PNS interface technologies [34]. A fully printed, flexible fingerprint-like three-axis tactile force and temperature sensor array was developed by Harada et al. [35] to measure simultaneously the tactile and slip force and temperature. He et al. [36] developed a flexible, self-powered, and self-clean T-ZNO/polyvinylidene fluoride (PVDF)/fabric multifunctional electronic skin. Li et al. designed a carbon nanotube (CNT)-based flexible skin-inspired sensory array for fingertips using silk-screen printing technique. The skin relies on three sensing materials and it can measure pressure, temperature, and humidity. Yogeswaran et al. fabricated a resistance e-skin flexible elastic tactile sensor for measurement of minimum pressure of 500 Pa, widely used in artificial robots and medical prosthesis, among others. Recently, Núñez et al. [37] developed a transparent tactile e-skin along with a single-layer graphene and coplanar interdigitated capacitive electrodes [37]. They also demonstrated the feasibility of large-scale and low-cost fabrication of a flexible and transparent e-skin for pressure sensing on a prosthetic hand [37]. Further, Osborn et al. [38] created a multilayered electronic dermis (e-dermis) that mimics the behavior of mechanoreceptors and nociceptors to deliver neuromorphic tactile and pain information to an amputee. A group in the École polytechnique fédérale de Lausanne (EPFL) developed an artificial skin that could help in rehabilitation and enhance virtual reality. The skin system contains soft sensors and pneumatic actuators which enable the artificial skin to conform to the exact shape of the wearer's wrist, e.g. and provide haptic feedback in the form of pressure and vibration. The sensor layer on top of the membrane layer contains soft electrodes made of a liquid–solid gallium mixture. These electrodes measure the skin deformation continuously and send the data to a microcontroller, which uses this feedback to fine-tune the sensation transmitted to the wearer in response to the wearer's movements and changes in external factors [39].

Table 2.2 presents a timeline summary of the evolution of tactile e-skin with special focus on robotic and prosthetic applications. It highlights in chronological order the development of tactile sensing from 1970 until 2010, and from 2010 until now, respectively. Along the aforementioned, our research group has been addressing this topic roughly for the past 10 years in a holistic way, managing the seamless design and implementation of the mechanical and electronic systems of the e-skin. Figure 2.3 depicts our proposed approach and application scenario: embedded, artificial distributed sensing and stimulation in prosthetic systems that provide high-fidelity, high-bandwidth tactile feedback to the prosthetic user.

Table 2.2 Timeline evolution of the tactile e-skin from 1970 until 2020

Years	Highlights and Achievements	Refs.
1974	Prosthetic hand with sensors	[21]
1980	Fictional Inspiration from	[22]
1984	*Star Wars* film/*Terminator* film	[79]
1982	1st Touchscreen computer from HP-150	[22, 80]
1984	General Electric (GE), first sensitive skin for a robotic arm based on infrared sensors	[22]
1999	Lumelsky et al. built a sensitive skin prototype module	[23]
2000	ASIMO Robot from Honda with tactile sensors	[20]
2001	Organic thin-film transistor as humidity sensor	[24]
2003	SNAVE intelligent anthropomorphic hand prosthesis contains force and slip sensors in the fingers	[81]
2004	Flexible active-matrix e-skin	[82]
2005	Conformable, stretchable, flexible large-area transparent e-skin based on an organic semiconductor	[27]
2006	Anthropomorphic cybernetic hand	[84]
2007	Wireless electronics, POSFET electronics, and transducers	[22]
2008	BioTAC human-like tactile sensors by SynTouch, LLC (USA)	[70]
2008	Dexterous shadow robotic arm endowed with BioTac SP sensors on fingertips	[70]
2009	Stretchable active-matrix OLEDs (organic light-emitting diode display) using printable elastic conductors	[22]
2010	Biodegradable OFETs, rechargeable and stretchable batteries	[84, 22]
2010	Macroscale nanowire e-skin capable of monitoring applied pressure profiles with high spatial resolution	[85]
2010	Icub Robot IIT, sensing arrays: ROBOSKIN project developed a robotic skin to cover large areas of the robot body	[86]
2011	Epidermal skin	[87]
2011	Stretchable, transparent skin-like pressure and strain sensors based on transparent elastic films of carbon nanotubes	[33]
2012	Self-healing e-skin	[31]
2012	Interlocked e-skin for monitoring signals from human heartbeats	[88]
2012	Multimodal artificial skin tactile for NAO Humanoid Robot, Hex-O-Skin at the Institute for Cognitive Systems, TUM-Germany	[20]
2012	Conformable e-skin with 3D shape for fingertips, which match the mechanics of the epidermis	[89]
2013	Rechargeable, stretchable batteries with self-similar serpentine interconnects and integrated wireless recharging systems	[90]
2013	Flexible and comfortable tactile sensors using screen printed P(VDFTrFE) and multiwalled carbon nanotubes (MWCNT)/polydimethylsiloxane (PDMS) composites	[91]
2014	Cosmic Lab e-skin based on polyvinylidene fluoride (PVDF) made of 64 sensors on table	[58]

(Continued)

Table 2.2   Continued

Years	Highlights and Achievements	Refs.
2014	Multimodal smart e-skin for prosthetics based on stretchable silicon nanoribbon electronics	[34]
2015	Interactive color changing chameleon-inspired stretchable e-skin	[92]
2016	Textile pressure sensors	[101]
2016	Temporary tattoo to control smartphone Duo Skin	[93]
2017	Flexible, multifunctional e-skin based on zinc oxide nanorod	[94]
2017	Robotic hand with sense of heat with intrinsically stretchable rubbery sensors	[95]
2018	Nanowire FET based neural element for robotic tactile sensing skin	[96]
2018	Screen printed tactile sensing arrays for prosthetics based on PVDF polymers	[64]
2018	Soft, thin, light weighted skin-like electronic system integrated and wirelessly activated fully soft robots	[97]
2018	Neuromorphic e-dermis that enables pain sensation to the prosthetic hands	[38]
2018	3D-printed tactile sensor on 3D-printed prosthetic hand	[98]
2019	Self-sensing pneumatic actuator skin used in closed-loop haptic feedback	[39]
2019	Flexible tactile e-skin sensor based on CNTs/ PDMS nanocomposites	[99]
2020	Closed loop 4D printed soft robot	[100]

Figure 2.3   Application scenario for restoring the sense of touch in prosthetics.

2.4 E-Skin Systems

E-skin is a hybrid stackwise arrangement (i.e. patches) that integrates both tactile sensing and touch interpretation units. It changes in dimension according to the target application (e.g. small patch for fingertips or large area for robotics or prosthetics). An e-skin is composed of a protective layer, a sensor layer, a signal-processing layer, and a substrate [2]. The polymer-based protective layer (e.g. polydimethylsiloxane [PDMS]) protects the sensor array and transfers the contact mechanical stimulus to the sensor array when the skin is touched. In turn, the sensor array converts the pressure information

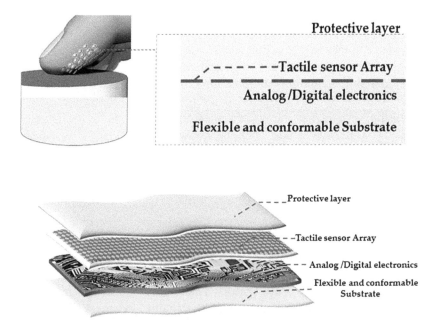

Figure 2.4 Illustration of the e-skin's general structure. (a) Frontal section of the e-skin of the different layers that would form the e-skin patch. (b) 3D image of the e-skin layers.

into electrical signals, which are further acquired and processed by the signal processing layer and then transmitted to the bottom structural material layer (i.e. the substrate). In addition to the flexibility, that allows the conformation of the skin onto various curved surfaces/shapes and the stretchability to support joint movement, the overall structure should be able to simultaneously sense different physical stimuli (including strain, twist, temperature, and humidity) with high sensitivity and distinguish them with efficient temporal and spatial resolutions. For instance, to emulate the human skin in terms of touch/pressure sensitivity, the e-skin should be able to recognize both medium-(10–100 KPa) and low pressures (<10 KPa) [3, 40].

Figure 2.4 illustrates an example of the general structure of the e-skin (adapted from Ref. [41]). Briefly, the e-skin system should comprise three main compartments, which are the sensing arrays that would substitute the sense of touch, interface electronic to convert analogue to digital tactile signals, and tactile data processing and decoding system [11]. The following sections will present the various tactile sensing transduction techniques and their respective research developments and application in the field of e-skin.

2.4.1 Transduction Mechanisms

So far, the development of the e-skin technology focused mainly on mimicking aspects of the human skin. The measured characteristics of touch are not only restricted to force and pressure, but also stiffness, texture, temperature, and strain. Transduction is considered as the first step to realize an artificial tactile sensing system that mimics the electrical output of the biological receptors [1]. This section describes the available tactile sensors used to transduce stimuli in the e-skins by converting them into electrical signals, namely, resistive sensors (such as strain gauges and piezoresistors), capacitive sensors, piezoelectric sensors, and optical sensors. In addition, this section will present a survey of the state-of-the-art distributed sensing and highlight limitations and challenges. Finally, the design requirements for tactile sensing system in prosthetics are depicted.

(a) Resistive

Piezoresistive sensors are resistive sensors. Their resistance changes upon the deformation formed by the applied mechanical stimulus. Piezoresistive tactile sensors are fabricated using conducting elastomers sandwiched between two vertically aligned electrodes or deposited between a pair of parallel inter-digitated electrodes. Piezoresistive strain gauges are considered the most common type of the referred sensors, whose resistance (R) depends on their geometry, which is calculated as

$$R = \rho L / A \qquad (2.1)$$

where ρ is the resistivity, L is the length, and A is the area. Under external strain, the geometry changes (i.e. either L or A is dominant). The resistance change could be given by $\Delta R/R = (1 + 2\upsilon)\varepsilon + \Delta\rho/\rho$, where υ and ε are Poisson's ratio and strain, respectively. Since the measurement of resistance is straightforward, piezoresistive tactile sensors have friendly electronic interface. They are less susceptible to interference and exhibit good sensitivity [9]. Another advantage is the applicability in microelectromechanical systems (MEMSs) or integrated to printed circuit boards [42]. Despite the mentioned advantages, piezoresistors suffer from hysteresis, temperature sensitivity, fragility, rigidity, and high cost. In 2014, Zhang et al. [43] developed a kind of resistive pressure sensor by transferring microstructure onto silk with PDMS. Jorgovanovic et al. [44] presented the static and dynamic characterization of piezoresistive sensors used for detecting the positions of prosthetic finger joints.

The feasibility of wireless communication between sensors and a receiving device, to reduce wires, was also discussed. Kane et al. [45] proposed a piezoresistive stress sensor array with high spatial resolution comparable to human dermis ($\sim$300 μm). Because of its unique stability, these sensors have great potential in the fields of prosthesis and robot artificial intelligence. They exhibited high potential for dexterous manipulation applications. Various applications with piezoresistive tactile sensors can also be found in stress and force measurement [45], stiffness of soft tissues detection [46], fingertip sensing [47], etc.

A strain gauge is another type of resistive sensors. It is a device adhered on the surface of an object to measure the strain caused by external pressure. The resistance of the foil changes with the stress applied on it. Strain gauges are more suitable to measure dynamic strains rather than static ones. The smaller a strain gauge is, the higher the accuracy. Strain gauges also exhibit nonlinear response and the measured strain is the average strain over the gauge length. Sensors of smaller size are flexible and robust to be applied over dexterous surfaces, such as prostheses, robots, and medical devices [50]. A finger-mounted tactile sensor based on the strain gauge which presented a linear response, a wide force sensitivity of 0–100 N with a resolution of 0.3 N, and a low hysteresis of 1.7% was proposed by Da Silva et al. [49, 50].

(b) Capacitive

A capacitive sensor consists of two parallel conductive plates that are separated by a dielectric material. The capacitance (C) is given by

$$C = \epsilon_0 \epsilon_r \frac{A}{d} \tag{2.2}$$

where ε_0 is the free space permittivity, ε_r is the relative static permittivity of the dielectric layer between the plates, A is the area of the overlap of the two plates, and d is the distance between the two plates. All variables reported in the equation except ε_0 are sensitive to changes in strain. When force is applied, the capacitance between the layers varies with the reduced distance between plates and the deformation of the middle dielectric material as well. Capacitive sensors demonstrate high sensitivity, robustness, and a large dynamic range. Due to these characteristics, capacitive sensing is widely used in robotic tactile sensing. Recently, most studies focus on the design of the dielectric layer and electrode structure. For the dielectric layer, it is difficult to achieve high sensitivity of sensors because of a high Young's modulus of elastomer material, such as some elastomer dielectrics including

PDMS with a small Young's modulus of as low as 5 kPa [51]. Schwartz et al. [52] proposed a flexible pressure-sensitive polymer transistor using a microstructural PDMS dielectric layer. Capacitive tactile sensor arrays are integrated into a prosthetic hand, thumb, and finger [53], temperature sensitivity, and low-power consumption [54]: they can be used for both dynamic and static force measurements. Additionally, their sensitivity to noise leads to relatively complex electronics for noise filtration. Capacitive sensors are considered as effective sensing elements and have been applied to multiaxis force measurement for gripping and objects manipulation, texture recognition [54], touch screen application [55], etc. A capacitive sensor for shear sensing was proposed with a size of 4 N [9]. It showed a high repeatability and approximately linear output within ±2 N; however, its dimension (3.5 mm × 1.6 mm × 1.6 mm) was a point to be considered in practical applications. Another capacitive tactile sensor was presented for gripping force measurement with a sensor range of 0–3000 mN [52].

(c) Piezoelectric

Piezoelectricity designates the ability of certain materials to generate an electrical charge in response to external mechanical stress. A piezoelectric tactile sensor is a device that is based on the direct piezoelectric effect: it measures changes generated due to external pressure. The generated electrical charge density is described by Equation (2.3):

$$D_i = d_{ijk}X_{jk} \tag{2.3}$$

where d_{ijk} is the piezoresistive coefficient of the material, X_{jk} is the external applied effect, and D_i is the charge density generated in the ith direction. Piezoelectric sensing is one of the few sensing techniques that do not require power supply; further, it also exhibits high sensitivity, reliability, and fast dynamic response. A wide response range of 0 to 1 kHz enables it to be a good choice for measurement of vibrations [56]. However, piezoelectric sensors are unsuitable for measuring static force and show low spatial resolution and poor temperature stability [56]. One of the most widely used piezoelectric materials to replicate the dynamic tactile mechanoreceptors in human fingertips is the PVDF. PVDF is a semicrystalline polymer consisting of long chain molecules with repeated units of CF-CH. Its strong piezoelectricity is attributed to the high electronegativity of fluoride atoms when compared with carbon atoms, which leads to a large dipole moment [57]. PVDF has many advantages, such as, mechanical flexibility, dimensional stability, high piezoelectric coefficients, low weight, formability into very thin sheets (5 μm), and relatively low price. Valle and his coworkers developed arrays of piezoelectric polymer transducers that can be used for large-area implementation of flexible artificial skin in a frequency range of 1 Hz–1 kHz [58]. To decrease the cross talk, affecting piezoelectric tactile sensors when presented in array form, dome-shaped cells of sensors have been proposed by Kim et al. [59].

Another promising piezoelectric material is zinc oxide (ZnO) nanotransducer because of its high flexibility and biocompatibility [62]. Its ability to generate electrical power when subjected to mechanical vibration leads to various potential applications, including wearable and self-power medical devices [61]. ZnO is a good candidate material for pressure and temperature sensors. During the past years, piezoelectric sensors have been used in prosthetic hands for the detection of slip [60], texture [62], and stiffness [63].

Besides the emergence of new piezoelectric materials, including poly (vinylidenefluoride-trifluoroethylene) (P(VDF-TrFE)), barium titanate (BaTiO$_3$), lead zirconate-titanate (PZT), and ZnO, has brought a turnaround for their development, replacing the conventional brittle ceramics and quartz. Flexible P(VDF-TrFE) is a suitable piezoelectric material due to its favorable chemical inertia, simple manufacturing, and large piezoelectric coefficient. Recently, Hoda et al. proposed screen-printed sensing arrays based on P(VDF-TrFE) piezoelectric polymers for prosthetic application [64]. They proved the reliability of the used technology and designed optimized skin patches to be mounted on the Michelangelo prosthetic hand from Ottobock and sensorized assistive glove [65].

(d) Other transduction mechanisms

An optical fiber force sensor mainly consists of a light source, a transduction medium, and an optical detector, which is often a photodiode. The light generated by the light source, usually light emitting diodes (LEDs), passes through the transduction medium which includes optical fibers and a modulator, and finally reaches the detector [66]. The detector circuit converts the light signal into an electrical one which is further processed. The intensity or the spectrum of the modulated light changes according to the variation of the applied force, which is the working principle of optical sensors. This major advantage enables optical sensors to be used in minimally invasive surgeries (MISs), where magnetic resonance imaging (MRI) procedures are widely used to provide high-quality images of living organs [66]. Optical sensors have a simple and compact structure, and high spatial resolution [63]. Most optical fibers are fragile and not as flexible as electric wires due to their relatively large size, which is another problem to be considered for dexterous hand applications.

2.4.2 Tactile Sensing Applications: Robotic and Prosthetic Hands

Among the numerous applications of tactile sensing systems, prosthetics and robotics have gained high relevance due to their important role in industrial applications and in daily life routines. Most studies focus on contact force or pressure to prevent slip to achieve a reliable grasp. Researchers have started employing tactile sensing technologies to provide sensory feedback to the users in prosthetics hands and arms to provide tactile feedback to amputees to allow them to regain lost functionality and reduce phantom limb pain. Sensory feedback would help patients to feel the prosthetic limb as a part of the body and relieve the stress coupled with high visual and auditory cognitive load arising from the control of prosthetics and robotic limbs. High performance tactile sensing is still required for ameliorating the operation speed and effectiveness of the current technologies. This section presents the commercial tactile sensors used in robotic hands and their limitations, shows the design requirements for tactile sensing systems for restoration of touch in prosthetics, and also includes a short survey of the tactile sensors in prosthetic hands (Table 2.3).

2.4.2.1 Tactile sensors in commercial robotic hands

Tactile sensors have been used by robotic hands, such as force, position, humidity, temperature, etc. to obtain stable grasp control, prescise object manipulation, and safer interaction with the surronding enviroment. An overview of sensors for robotic hands is presented in Ref. [68]. Ueda et al. [69] developed a four fingered actuated NAIST hand to test grasping with vison-based tactile sensors by estimating slip margin. Choi et al. present a PVDF fingertip tactile sensor, which detects normal contact force and slip. Wettels et al. depict three sensing modalities (i.e. force, vibration, and thermal sensing) of the developed BioTac® biomimetic tactile sensor sensor [70]. Vibration and thermal sensations are used to discriminate between object properties. It has been integrated into the ShadowHand by Xu et al. and used for identification of objects through tactile data [71]. In the "Roboskin" project, Cannata et al. developed an artifical skin and applied it onto different robotic platforms (such as iCub and Schunk robot hand) to accomplish efficient and safe human–robot interaction [72].

Table 2.3 Characteristics of different tactile sensors used for e-skin fabrication

Sensing Principle	Operation Basis	Trade-Offs	
		Pros	Cons
Piezoresistivity	Resistance varies with the deformation caused by applied force.	Simple electronics High sensitivity Ease of integrating in MEMS Compatible with VLSI Resistant to interference Low cost	Hysteresis Temperature Sensitivity Fragile and rigid Lack of reproducibility High power consumption
Capacitive	Capacitance varies with the deformation caused by applied force.	Sensitivity of small force change Reliability Large dynamic range suitable for both dynamic and static force measurements Low temperature sensitivity Low power consumption	Limited spatial resolution Noise sensitivity Complex electronics Cross talk between elements Hysteresis
Piezoelectricity	Electric charge is produced when a force is applied on it.	No need for power supply High reliability Fast dynamic response High sensitivity High accuracy	Low spatial resolution High temperature sensitivity Inability to sense static value
Optical	The intensity or the spectrum of light varies with the applied force	Immune to electromagnetic fields High spatial resolution Wide sensing range Good reliability	Fragile and rigid Large size Inability to transparency and highly reflective surface

Emphasis toward developing robotic hands that provide sensor feedback to the human operator motivated the development of wide variety of commercial sensor systems, e.g. BioTac® [70], DigiTacts [73], Tekscan^{TM} [74], Weiss [75], and Peratech [76]. Even though they are advanced technologies, they still show several limitations that restrict their applicability in systems that require flexible, compact, robust, and power efficient, e.g. prosthetics. In this regard, the following are the limitations: high power consumption (Weiss 250 mW), narrow applicability and high cost (e.g. BioTac® only available for fingertips, low resolution (DigiTacts, 22 taxels employed on Allegro robotic hand), low frame rate (Weiss 400 f/s, Tekscan^{TM} 200 f/s, DigiTacts 100 f/s), difficult system integration (TekscanTM complex wiring), and large sensor array size and/or thickness (Peratech: large size, e.g. 15×36 cm^2, Weiss: large thickness, i.e. 2 cm for fingertips).

2.4.2.2 Tactile sensory systems in prosthetics hands

Emulating the biological sense of touch for hand prostheses mainly depends on the development of the following: (a) an articulated tactile sensing system that includes tactile sensors and/or sensors that measure pain, temperature, and proprioception [3, 77], (b) an adaptive sensory signal encoding algorithm, and (c) signal transmission and transduction methods to convey the sensory information to the nervous system. Prosthetic hands, if at all sensorized, typically possess two kinds of sensors: position sensors to provide the hand with proprioceptive information and force/tactile sensors for estimating mechanical interactions with the environment. Still, the measured character-istics cannot be limited to force or position only; ideally, they should also include stiffness, texture, shape, etc. Osborn et al. [38] have used them in the development of a closed loop upper limb prosthetic system that measures contact force to detect object slippage and reduce grip strength to prevent breaking of the objects. Other tactile sensors use piezoelectricity, and they are usually employed to measure dynamic forces/pressures. One of the examples is the modular prosthetic limb developed by researchers from Johns Hopkins University, and this anthropomorphic prosthetic hand based on 100 sensors could provide high resolution tactile- and position-sensing capabilities [77]. Southampton hand used two different types of sensors on the fingertips to restore tactile information: piezoelectric sensors based on PZT for slippage detection and FSR sensors for measuring force [32]. For applications that require high sensitivity and resolution, capacitive sensors are considered the best candidates, as they can be used for both static and dynamic force measurements. A capacitive based tactile sensor has been utilized for mea-suring the gripping force in the range of 0–3000 mN, and it has been tested on the prosthetic hand. Finally, looking on the commercial prosthetics hands, only the Sensor Hand from Otto Bock is provided with a slippage detecting system (i.e. SUVA sensor system), while recent models such as i-Limb, the Bebionic, are not yet provided with force or tactile feedback [8, 11, 32].

Today's tactile sensing systems encounter many challenges that limit their integration in prosthetic systems, such as designs issues, spatial distribution, low signal-to-noise ratio (SNR), cross talk, wireless communication, and the lack of signal processing and machine learning methods to encode the acquired data from tactile sensors [40, 41].

2.5 Requirements and Challenges

Although various types of wearable e-skins have achieved flexibility, elasticity, sensitivity, and wide range by adopting novel materials, fabrication techniques, and sensing mechanisms, there are still challenges to overcome, to develop the e-skin that is on par with the human skin at performance and practical application. For instance, such limitations include the integration of large number of sensing elements in limited space especially at fingertips and sensor-addressing problems (i.e. how to read data from each discrete sensor). E-skins for intelligent robots and prostheses should possess the ability to perceive and differentiate multidirectional forces such as normal, tensile, shear, and bending forces without cross talk. Furthermore, multiple important factors are still needed to be addressed such as signal processing, power supply, integration, and biocompatibility especially for biomedical applications.

Sensors for prosthetic hands transduce various modalities of tactile stimuli aiming at recreating naturalistic perception. It is expected that artificial tactile sensors demonstrate small spatial resolution (≤ 1 mm for fingertips, 5 mm for hand and palm, 20–30 mm, e.g. limbs, torso, etc.), high sensitivity varying from 0.01 to 10 N, which extend along the tactile frequency range (<1 Hz–1 kHz), low hysteresis, fast and linear response (<1 ms), wide dynamic range, and high reliability. Furthermore, it needs to exhibit high electromechanical bandwidth to detect fast events (e.g. incipient slip), large force/pressure (e.g. 1–1000 g) for daily activities, adequate size and pitch (e.g. 1 cm/1.5–2 mm for fingertips), and customizable shape of the e-skin patches and sensor number. Additionally, low cost, low power consumption, and scalability are the major factors for the prosthetic application. The main design requirements of tactile sensing system in prosthetics are summarized in Table 2.4.

Herein, this section summarizes the e-skin requirements:

- Linear sensitivity over large pressure range would guarantee an accurate tactile sensor output information without complex signal processing; however, it would lower power consumption and pave the way for miniaturization.

Table 2.4   Design requirements for tactile sensing system in prosthetics

Design Benchmarks	Specifications
Dynamic force range	0.01 N–10 N
Spatial resolution	$\leq$1 mm for small-sensing areas (e.g. fingertips)
Taxel for a sening element/pitch for sensing arrays pitch	$\geq$5mm $\div$ 20–30 mm for large-sensing arrays (e.g. limbs, torso, etc.)
Sensor frequency bandwidth	<1 Hz–1 kHz
Sensor response time	About 1 kHz (1 ms)
Sensing detection ability	Normal and shear forces; vibrations
Mechanical attributes	Flexible, stretchable, conformable and soft, robust and durable
Electrical attributes	Low power, minimal wiring and cross talk, electrically and magnetically minimal sensitivity
Sensor response	Monotonic, fast and linear, stable and repeatable, low hysteresis
Temporal variation	Both dynamic and static

- A self-powered system could generate or collect electrical power from various sources such as body motion or environmental sources, to obtain a standalone device.
- Flexibility, conformability, and stretchability for large-area sensing for robotics or prostheses.
- Biocompatibility and biodegradability of materials will reduce immune reaction in biomedical applications.

2.6 Conclusion and Perspectives

Given the importance of tactile sensing in the exploration of the environment and the daily life interactions, researchers are still striving to understand the sense of touch and aim to develop smart tactile sensing systems that could mimic its characteristics and functions. E-skin is an artificial skin that aims to replicate the human skin. It could be fabricated using several transduction techniques, materials and structural designs, depending on the target application. E-skin should be flexible and stretchable; additionally, it must have multifunctional sensing capabilities and cover large areas at low cost. E-skin has been included in numerous applications especially in robotics; however, its implementation into prosthetics is still in its infancy due to the technical difficulties and complicated nature of the human tactile sensation. High potential relies on the fabrication and the development of flexible, stretchable, and robust large-area multifunctional intelligent e-skins that respond to external physical stimuli and provide better temporal and spatial resolutions.

Combining the e-skin with the next generation technologies such as Internet of things (IOT), artificial intelligence (AI), and virtual reality (VR) could pave the way for innovative future applications.

The construction of wireless communication devices assists the remote monitoring/control and data collection from multiple sensors, which would enable real-time and personalized applications. The perception of various signals and environments based on deep learning and machine learning through pattern recognition and modeling will be an essential factor for artificial intelligent robots and advanced wearables. Displaying tactile information by visualization techniques may enhance haptic interfaces and VR applications. Conclusively, combining these elements (i.e. signal transmission, perception, and tactile sensing information display) with the e-skins will provide technological innovation.

References

[1] Dahiya, R.S., Metta, G., Valle, M. and Sandini, G., 2009. Tactile sensing – from humans to humanoids. IEEE Transactions on Robotics, 26(1), pp. 20. DOI: 10.1109/TRO.2009.2033627

[2] Zou, L., Ge, C., Wang, Z.J., Cretu, E. and Li, X., 2017. Novel tactile sensor technology and smart tactile sensing systems: A review. Sensors, 17(11), p. 2653. DOI: 10.3390/s17112653

[3] Chortos, A., Liu, J. and Bao, Z., 2016. Pursuing prosthetic electronic skin. Nature Materials, 15(9), pp. 937–50. DOI: 10.1038/nmat4671

[4] Ciancio, A.L., Cordella, F., Barone, R., Romeo, R.A., Bellingegni, A.D., Sacchetti, R., Davalli, A., Di Pino, G., Ranieri, F., Di Lazzaro, V. and Guglielmelli, E., 2016. Control of prosthetic hands via the peripheral nervous system. Frontiers in Neuroscience, 10, p. 116. DOI: 10.3389/fnins.2016.00116

[5] Silvera-Tawil, D., Rye, D. and Velonaki, M., 2015. Artificial skin and tactile sensing for socially interactive robots: A review.

Robotics and Autonomous Systems, 63, pp. 230–243. DOI: 10.1016/j.robot.2014.09.008

[6] Girão, P.S., Ramos, P.M.P., Postolache, O. and Pereira, J.M.D., 2013. Tactile sensors for robotic applications. Measurement, 46(3), pp. 1257–1271. DOI: 10.1016/j.measurement.2012.11.015

[7] Kumar, V. and Todorov, E., November, 2015. "MuJoCo HAPTIX: A virtual reality system for hand manipulation," 2015 IEEE-RAS 15th International Conference on Humanoid Robots (Humanoids), Seoul, 2015, pp. 657–663. DOI: 10.1109/HUMANOIDS.2015.7363441

[8] Chi, C., Sun, X., Xue, N., Li, T. and Liu, C., 2018. Recent progress in technologies for tactile sensors. Sensors, 18(4), p. 948. DOI:10.3390/s18040948

[9] Tiwana, M.I., Redmond, S.J. and Lovell, N.H., 2012. A review of tactile sensing technologies with applications in biomedical engineering. Sensors and Actuators A: Physical, 179, pp. 17–31. DOI: 10.1016/j.sna.2012.02.051

[10] Ly, H.H., Tanaka, Y., Fukuda, T. et al., 2017. Grasper having tactile sensing function using acoustic reflection for laparoscopic surgery. International Journal of Computer Assisted Radiology and Surgery, 12, pp. 1333–1343. DOI:10.1007/s11548–017-1592-7

[11] Almansoori, M.T., Li, X. and Zheng, L., 2019. A brief review on E-skin and its multifunctional sensing applications. Current Smart Materials, 4(1), pp. 3–14. DOI: 10.2174/2405465804666190313154903

[12] Klatzky, R.L. and Lederman, S.J., 2003. Touch. In A.F. Healy and R.W. Proctor (Eds.). Handbook of Psychology: Experimental Psychology, Vol. 4 (pp. 147–176). John Wiley & Sons Inc. DOI: 10.1002/0471264385.wei0406

[13] Darian-Smith, I., 2011. The sense of touch: performance and peripheral neural processes. Comprehensive Physiology, pp. 739–788. DOI: 10.1002/cphy.cp010317

[14] Graziano, M.S.A. and Botvinick, M.M., 2002. How the brain represents the body: Insights from neurophysiology and psychology. In W. Prinz and B. Hommel (Ed.). Common Mechanisms in Perception and Action: Attention and Performance XIX, pp. 136–157. Oxford University Press.

[15] Johansson, R.S. and Westling, G., 1984. Roles of glabrous skin receptors and sensorimotor memory in automatic control of precision grip when lifting rougher or more slippery objects. Experimental Brain Research, 56(3), pp. 550–564. DOI: 10.1007/BF00237997

[16] Vallbo, A.B. and Johansson, R.S., 1984. Properties of cutaneous mechanoreceptors in the human hand related to touch sensation. Human Neurobiology, 3(1), pp. 3–14.

[17] Johansson, R.S. and Flanagan, J.R., 2009. Coding and use of tactile signals from the fingertips in object manipulation tasks. Nature Reviews Neuroscience, 10(5), pp. 345–359. DOI: 10.1038/nrn2621

[18] Park, M., Bok, B.G., Ahn, J.H. and Kim, M.S., 2018. Recent advances in tactile sensing technology. Micromachines, 9(7), p. 321. DOI: 10.3390/mi9070321

[19] Bensmaia, S.J., Leung, Y.Y., Hsiao, S.S. and Johnson, K.O., 2005. Vibratory adaptation of cutaneous mechanoreceptive afferents. Journal of Neurophysiology, 94(5), pp. 3023–3036. DOI: 10.1152/jn.00002.2005

[20] Dahiya, R., Yogeswaran, N., Liu, F., Manjakkal, L., Burdet, E., Hayward, V. and Jörntell, H., 2019. Large-area soft e-Skin: The challenges beyond sensor designs. Proceedings of the IEEE, 107(10), pp. 2016–2033. DOI: 10.1109/JPROC.2019.2941366

[21] Clippinger, F.W., Avery, R. and Titus, B.R., 1974. A sensory feedback system for an upper-limb amputation prosthesis. Bulletin of Prosthetic Research, 10, p. 247.

[22] Hammock, M.L., Chortos, A., Tee, B.C.K., Tok, J.B.H. and Bao, Z., 2013. 25th anniversary article: The evolution of electronic skin (e-skin): A brief history, design considerations, and recent progress. Advanced Materials, 25(42), pp. 5997–6038. DOI: 10.1002/adma.201302240

[23] Lumelsky, V.J., Shur, M.S. and Wagner, S., 2001. Sensitive skin. IEEE Sensors Journal, 1(1), pp. 41–51. DOI: 10.1142/9789812792525_0004

[24] Torsi, L., Dodabalapur, A., Cioffi, N., Sabbatini, L. and Zambonin, P.G., 2001. NTCDA organic thin-film-transistor as humidity sensor: Weaknesses and strengths. Sensors and Actuators B: Chemical, 77 (1–2), pp. 7–11. DOI: 10.1016/S0925–4005(01)00664-5

[25] Lacour, S.P., Wagner, S., Huang, Z. and Suo, Z., 2003. Stretchable gold conductors on elastomeric substrates. Applied Physics Letters, 82(15), pp. 2404–2406. DOI: 10.1063/1.1565683

[26] Kim, D.H., Song, J., Choi, W.M., Kim, H.S., Kim, R.H., Liu, Z., Huang, Y.Y., Hwang, K.C., Zhang, Y.W. and Rogers, J.A., 2008. Materials and noncoplanar mesh designs for integrated circuits with linear elastic responses to extreme mechanical deformations. Proceedings of

the National Academy of Sciences, 105(48), pp. 18675–18680. DOI 10.1073/pnas.0807476105

[27] Someya, T., Kato, Y., Sekitani, T., Iba, S., Noguchi, Y., Murase, Y., Kawaguchi, H. and Sakurai, T., 2005. Conformable, flexible, large-area networks of pressure and thermal sensors with organic transistor active matrixes. Proceedings of the National Academy of Sciences, 102(35), pp. 12321–12325. DOI: 10.1073/pnas.0502392102

[28] Metzger, C., Fleisch, E., Meyer, J., Dansachmüller, M., Graz, I., Kaltenbrunner, M., Keplinger, C., Schwödiauer, R. and Bauer, S., 2008. Flexible-foam-based capacitive sensor arrays for object detection at low cost. Applied Physics Letters, 92(1), p. 013506. DOI: 10.1063/1.2830815

[29] Graz, I., Kaltenbrunner, M., Keplinger, C., Schwödiauer, R., Bauer, S., Lacour, S.P. and Wagner, S., 2006. Flexible ferroelectret field-effect transistor for large-area sensor skins and microphones. Applied Physics Letters, 89(7), p. 073501. DOI: 10.1063/1.2335838

[30] Mannsfeld, S.C., Tee, B.C., Stoltenberg, R.M., Chen, C.V.H., Barman, S., Muir, B.V., Sokolov, A.N., Reese, C. and Bao, Z., 2010. Highly sensitive flexible pressure sensors with microstructured rubber dielectric layers. Nature Materials, 9(10), pp. 859–864. DOI: 10.1038/nmat2834

[31] Tee, B.C., Wang, C., Allen, R. and Bao, Z., 2012. An electrically and mechanically self-healing composite with pressure-and flexion-sensitive properties for electronic skin applications. Nature Nanotechnology, 7(12), pp. 825–832. DOI:10.1038/nnano.2012.192

[32] Yu, Z., Niu, X., Liu, Z. and Pei, Q., 2011. Intrinsically stretchable polymer light-emitting devices using carbon nanotube-polymer composite electrodes. Advanced Materials, 23(34), pp. 3989–3994. DOI: 10.1002/adma.201101986

[33] Lipomi, D.J., Tee, B.C.K., Vosgueritchian, M. and Bao, Z., 2011. Stretchable organic solar cells. Advanced Materials, 23(15), pp. 1771–1775. DOI: 10.1002/adma.201004426

[34] Kim, J., Lee, M., Shim, H.J., Ghaffari, R., Cho, H.R., Son, D., Jung, Y.H., Soh, M., Choi, C., Jung, S. and Chu, K., 2014. Stretchable silicon nanoribbon electronics for skin prosthesis. Nature Communications, 5(1), pp. 1–11. DOI: 10.1038/ncomms6747

[35] Harada, S., Kanao, K., Yamamoto, Y., Arie, T., Akita, S. and Takei, K., 2014. Fully printed flexible fingerprint-like three-axis tactile and slip force and temperature sensors for artificial skin. ACS Nano, 8(12), pp. 12851–12857. DOI: 10.1021/nn506293y

[36] He, H., Fu, Y., Zang, W., Wang, Q., Xing, L., Zhang, Y. and Xue, X., 2017. A flexible self-powered T-ZnO/PVDF/fabric electronic-skin with multi-functions of tactile-perception, atmosphere-detection and self-clean. Nano Energy, 31, pp. 37–48. DOI: 10.1016/j.nanoen.2016.11.020

[37] Núñez, C.G., Navaraj, W.T., Polat, E.O. and Dahiya, R., 2017. Energy-autonomous, flexible, and transparent tactile skin. Advanced Functional Materials, 27(18), p. 1606287. DOI: 10.1002/adfm.20160 6287

[38] Osborn, L.E., Dragomir, A., Betthauser, J.L., Hunt, C.L., Nguyen, H.H., Kaliki, R.R. and Thakor, N.V., 2018. Prosthesis with neuro-morphic multilayered e-dermis perceives touch and pain. Science Robotics, 3(19), p. eaat3818. DOI: 10.1126/scirobotics.aat3818

[39] Sonar, H.A., Gerratt, A.P., Lacour, S.P. and Paik, J., 2020. Closed-loop haptic feedback control using a self-sensing soft pneumatic actuator skin. Soft Robotics, 7(1), pp. 22–29, DOI: 10.1089/soro.2019.0013

[40] H. Fares et al., "Distributed Sensing and Stimulation Systems for Sense of Touch Restoration in Prosthetics," *2017 New Generation of CAS (NGCAS)*, Genova, 2017, pp. 177–180. DOI: 10.1109/NGCAS.2017.54

[41] Ibrahim, A., Pinna, L., Seminara, L. and Valle, M., 2017. Achievements and Open Issues Toward Embedding Tactile Sensing and Interpretation into Electronic Skin Systems. Material-Integrated Intelligent Systems-Technology and Applications; John Wiley & Sons, Ltd.: WeinHeim, Germany, pp. 571–594. DOI:10.1002/9783527679249. ch23.

[42] Stassi, S., Cauda, V., Canavese, G. and Pirri, C.F., 2014. Flexible tactile sensing based on piezoresistive composites: A review. Sensors, 14(3), pp. 5296–5332. DOI: 10.3390/s140305296

[43] Wang, X., Gu, Y., Xiong, Z., Cui, Z. and Zhang, T., 2014. Silk-molded flexible, ultrasensitive, and highly stable electronic skin for monitoring human physiological signals. Advanced Materials, 26(9), pp. 1336–1342. DOI: 10.1002/adma.201304248

[44] Jorgovanovic, N., Dosen, S., Djozic, D.J., Krajoski, G. and Farina, D., 2014. Virtual grasping: Closed-loop force control using electrotactile feedback. Computational and Mathematical Methods in Medicine, pp. 1–13., DOI: 10.1155/2014/120357

[45] Kane, B.J., Cutkosky, M.R. and Kovacs, G.T., 2000. A traction stress sensor array for use in high-resolution robotic tactile imaging. Journal of Microelectromechanical Systems, 9(4), pp. 425–434.

[46] Kalantari, M., Ramezanifard, M., Ahmadi, R., Dargahi, J. and Kövecses, J., 2011. A piezoresistive tactile sensor for tissue characterization during catheter-based cardiac surgery. The International Journal of Medical Robotics and Computer Assisted Surgery, 7(4), pp. 431–440. DOI: 10.1002/rcs.413

[47] Koiva, R., Zenker, M., Schürmann, C., Haschke, R. and Ritter, H.J., July, 2013. A highly sensitive 3D-shaped tactile sensor. In 2013 Proceedings of the IEEE ASME International Conference on Advanced Intelligent Mechatronics (pp. 1084–1089). DOI: 10.1109/AIM.2013.6584238

[48] Engel, J., Chen, J. and Liu, C., 2003. Development of polyimide flexible tactile sensor skin. Journal of Micromechanics and Microengineering, 13(3), pp. 359–366. DOI: 10.1088/0960–1317/13/3/302

[49] Da Silva, J.G., de Carvalho, A.A. and da Silva, D.D., 2002. A strain gauge tactile sensor for finger-mounted applications. IEEE Transactions on Instrumentation and Measurement, 51(1), pp. 18–22. DOI: 10.1109/19.989890

[50] Li, K., Fang, Y., Zhou, Y. and Liu, H., 2017. Non-invasive stimulation-based tactile sensation for upper-extremity prosthesis: A review. IEEE Sensors Journal, 17(9), pp. 2625–2635. DOI: 10.1109/JSEN.2017.2674965

[51] Lei, K.F., Lee, K.F. and Lee, M.Y., 2012. Development of a flexible PDMS capacitive pressure sensor for plantar pressure measurement. Microelectronic Engineering, 99(C), pp. 1–5. DOI: 10.1016/j.mee.2012.06.005

[52] Schwartz, G., Tee, B.C.K., Mei, J., Appleton, A.L., Kim, D.H., Wang, H. and Bao, Z., 2013. Flexible polymer transistors with high pressure sensitivity for application in electronic skin and health monitoring. Nature Communications, 4(1), pp. 1–8. DOI: 10.1038/ncomms2832

[53] Wang, Y., Xi, K., Liang, G., Mei, M. and Chen, Z., July 2014. "A flexible capacitive tactile sensor array for prosthetic hand real-time contact force measurement," 2014 IEEE International Conference on Information and Automation (ICIA), Hailar, 2014, pp. 937–942. DOI: 10.1109/ICInfA.2014.6932786

[54] Muhammad, H.B., Recchiuto, C., Oddo, C.M., Beccai, L., Anthony, C.J., Adams, M.J., Carrozza, M.C. and Ward, M.C.L., 2011.

A capacitive tactile sensor array for surface texture discrimination. Microelectronic Engineering, 88(8), pp. 1811–1813. DOI: 10.1016/j.mee.2011.01.045

[55] Kim, H.K., Lee, S. and Yun, K.S., 2011. Capacitive tactile sensor array for touch screen application. Sensors and Actuators A: Physical, 165(1), pp. 2–7. DOI: 10.1016/j.sna.2009.12.031

[56] Seminara, L., Capurro, M., Cirillo, P., Cannata, G. and Valle, M., 2011. Electromechanical characterization of piezoelectric PVDF polymer films for tactile sensors in robotics applications. Sensors and Actuators A: Physical, 169(1), pp. 49–58. DOI: 10.1016/j.sna.2011.05.004

[57] Dargahi, J., 2000. A piezoelectric tactile sensor with three sensing elements for robotic, endoscopic and prosthetic applications. Sensors and Actuators A: Physical, 80(1), pp. 23–30. DOI: 10.1016/S0924–4247(99)00295–2

[58] Seminara, L., Pinna, L., Valle, M., Basiricò, L., Loi, A., Cosseddu, P., Bonfiglio, A., Ascia, A., Biso, M., Ansaldo, A. and Ricci, D., 2013. Piezoelectric polymer transducer arrays for flexible tactile sensors. IEEE Sensors Journal, 13(10), pp. 4022–4029. DOI: 10.1109/JSEN.2013.2268690

[59] Kim, M.S., Ahn, H.R., Lee, S., Kim, C. and Kim, Y.J., 2014. A dome-shaped piezoelectric tactile sensor arrays fabricated by an air inflation technique. Sensors and Actuators A: Physical, 212, pp. 151–158. DOI: 10.1016/j.sna.2014.02.023

[60] Cotton, D.P.J., Chappell, P.H., Cranny, A., White, N.M. and Beeby, S.P., 2007. A novel thick-film piezoelectric slip sensor for a prosthetic hand. IEEE Sensors Journal, 5(7), pp. 752–761. DOI: 10.1109/JSEN.2007.894912

[61] Dakua, I. and Afzulpurkar, N., 2013. Piezoelectric energy generation and harvesting at the nanoscale: Materials and devices. Nanomaterials and Nanotechnology, 3(Godište 2013), pp. 3–21. DOI: 10.5772/56941

[62] Takamuku, S., Gomez, G., Hosoda, K. and Pfeifer, R., October, 2007. "Haptic discrimination of material properties by a robotic hand," 2007 IEEE 6th International Conference on Development and Learning, London, 2007, pp. 1–6. DOI: 10.1109/DEVLRN.2007.4354057

[63] Omata, S. and Terunuma, Y., 1992. New tactile sensor like the human hand and its applications. Sensors and Actuators A: Physical, 35(1), pp. 9–15. DOI: 10.1016/0924–4247(92)87002-X

[64] Fares, H., Seminara, L., Pinna, L., Valle, M., Groten, J., Hammer, M., Zirkl, M. and Stadlober, B., May, 2018. "Screen Printed Tactile

Sensing Arrays for Prosthetic Applications," 2018 IEEE International Symposium on Circuits and Systems (ISCAS), Florence, 2018, pp. 1–4. DOI: 10.1109/ISCAS.2018.8351892

[65] Fares, H., Abbass, Y., Valle, M., and Seminara, L., 2020. Validation of screen-printed electronic skin based on piezoelectric polymer sensors. Sensors, 20(4): p. 1160. DOI 10.3390/s20041160

[66] Maekawa, H., Tanie, K. and Komoriya, K., October, 1993. "A finger-shaped tactile sensor using an optical waveguide," Proceedings of IEEE Systems Man and Cybernetics Conference – SMC, Le Touquet, France, 1993, pp. 403–408 vol. 5, DOI: 10.1109/ICSMC.1993.390885

[67] Ataollahi, A., Polygerinos, P., Puangmali, P., Seneviratne, L.D. and Althoefer, K., 2010. "Tactile sensor array using prismatic-tip optical fibers for dexterous robotic hands," 2010 IEEE/RSJ International Conference on Intelligent Robots and Systems, Taipei, 2010, pp. 910–915, DOI: 10.1109/IROS.2010.5649109.

[68] Saudabayev, A. and Varol, H.A., 2015. Sensors for robotic hands: A survey of state of the art. IEEE Access, 3, pp. 1765–1782. DOI: 10.1109/ACCESS.2015.2482543

[69] Ueda, J., Ishida, Y., Kondo, M. and Ogasawara, T., April, 2005. "Development of the NAIST-Hand with Vision-based Tactile Fingertip Sensor," Proceedings of the 2005 IEEE International Conference on Robotics and Automation, Barcelona, Spain, 2005, pp. 2332–2337, DOI: 10.1109/ROBOT.2005.1570461.

[70] "Robotics & Prosthetics – SynTouch." [Online]. Available at: https://syntouchinc.com/robotics/ [accessed 09-Apr-2020].

[71] Fishel, J.A. and Loeb, G.E., 2012. Bayesian exploration for intelligent identification of textures. Frontiers in Neurorobotics, 6, p. 4. DOI: 10.3389/fnbot.2012.00004

[72] Billard, A., Bonfiglio, A., Cannata, G., Cosseddu, P., Dahl, T., Dautenhahn, K., Mastrogiovanni, F., Metta, G., Natale, L., Robins, B. and Seminara, L., 2013. The roboskin project: Challenges and results. In Romansy 19–Robot Design, Dynamics and Control, pp. 351–358. Springer, Vienna. DOI: 10.1007/978–3–7091–1379-0_43

[73] "DigiTacts – PPS." [Online]. Available at: https://pressureprofile.com/sensors/digitacts [accessed 09-Apr-2020].

[74] "Sensors for Pressure Mapping and Force Measurement | Tekscan." [Online]. Available at: https://www.tekscan.com/products-solutions/sensors [accessed 09-Apr-2020].

[75] "Sensor fingers | WEISS ROBOTICS." [Online]. Available at: https://weiss-robotics.com/sensor-fingers/ [accessed 09-Apr-2020].

[76] "Peratech – QTC Multi-touch Sensors." [Online]. Available at: https://www.peratech.com/qtc-multi-touch-sensors/ [accessed 09-Apr-2020].

[77] Chortos, A. and Bao, Z., 2014. Skin-inspired electronic devices. Materials Today, 17(7), pp. 321–331. DOI: 10.1016/j.mattod.2014.05.006

[78] "Revolutionizing Prosthetics MPL." [Online]. Available at: https://www.jhuapl.edu/Prosthetics/ResearchMPL [accessed 09-Apr-2020].

[79] Kershner, I., Lucas, G., Brackett, L. and Kasdan, L., 2004. Star Wars episode 5: The empire strikes back 1980, 20th century Fox.

[80] Vintage Computing and Gaming homepage, www.vintagecomputing.com

[81] Warwick, K., Gasson, M., Hutt, B., Goodhew, I., Kyberd, P., Andrews, B., Teddy, P. and Shad, A., 2003. The application of implant technology for cybernetic systems. Archives of Neurology, 60(10), pp. 1369–1373. DOI:10.1001/archneur.60.10.1369

[82] Someya, T., Sekitani, T., Iba, S., Kato, Y., Kawaguchi, H. and Sakurai, T., 2004. A large-area, flexible pressure sensor matrix with organic field-effect transistors for artificial skin applications. Proceedings of the National Academy of Sciences, 101(27), pp. 9966–9970. DOI: 10.1073/pnas.0401918101

[83] Micera, S., Carrozza, M.C., Beccai, L., Vecchi, F. and Dario, P., 2006. Hybrid bionic systems for the replacement of hand function. Proceedings of the IEEE, 94(9), pp. 1752–1762.

[84] Kaltenbrunner, M., Kettlgruber, G., Siket, C., Schwödiauer, R. and Bauer, S., 2010. Stretchable batteries: Arrays of ultracompliant electrochemical dry gel cells for stretchable electronics (Adv. Mater. 18/2010). Advanced Materials, 22(18) DOI:10.1002/adma.201090061

[85] Takei, K., Takahashi, T., Ho, J.C., Ko, H., Gillies, A.G., Leu, P.W., Fearing, R.S. and Javey, A., 2010. Nanowire active-matrix circuitry for low-voltage macroscale artificial skin. Nature Materials, 9(10), pp. 821–826. DOI:/10.1038/nmat2835

[86] Billard, A., Bonfiglio, A., Cannata, G., Cosseddu, P., Dahl, T., Dautenhahn, K., Mastrogiovanni, F., Metta, G., Natale, L., Robins, B. and Seminara, L., 2013. The roboskin project: Challenges and results. In Romansy 19–Robot Design, Dynamics and Control (pp. 351–358). Springer. DOI: 10.1007/978-3-7091-1379-0_43

[87] Kim, D.H., Lu, N., Ma, R., Kim, Y.S., Kim, R.H., Wang, S., Wu, J., Won, S.M., Tao, H., Islam, A. and Yu, K.J., 2011. Epidermal electronics. Science, 333(6044), pp. 838–843. DOI:10.1126/science.1206157

[88] Pang, C., Lee, G., Kim, T. et al., 2012. A flexible and highly sensitive strain-gauge sensor using reversible interlocking of nanofibres. Nature Materials, 11, pp. 795–801. DOI: 10.1038/nmat3380

[89] Ying, M., Bonifas, A.P., Lu, N., Su, Y., Li, R., Cheng, H., Ameen, A., Huang, Y. and Rogers, J.A., 2012. Silicon nanomembranes for fingertip electronics. Nanotechnology, 23(34), p. 344004. DOI: 10.1088/0957-4484/23/34/344004

[90] Xu, S., Zhang, Y., Cho, J. et al., 2013. Stretchable batteries with self-similar serpentine interconnects and integrated wireless recharging systems. Nature Communications, 4, 1543. DOI: 10.1038/ncomms2553

[91] Khan, S., Dahiya, R., Tinku, S. and Lorenzelli, L., 2014. Conformable tactile sensing using screen printed P(VDF-TrFE) and MWCNT-PDMS composites. Sensors, IEEE, Valencia, pp. 862–865, DOI: 10.1109/ICSENS.2014.6985136.

[92] Chou, H.H., Nguyen, A., Chortos, A., To, J.W., Lu, C., Mei, J., Kurosawa, T., Bae, W.G., Tok, J.B.H. and Bao, Z., 2015. A chameleon-inspired stretchable electronic skin with interactive colour changing controlled by tactile sensing. Nature Communications, 6(1), pp. 1–10. DOI:/10.1038/ncomms9011

[93] Kao, H.L., Holz, C., Roseway, A., Calvo, A. and Schmandt, C., September, 2016. DuoSkin: rapidly prototyping on-skin user interfaces using skin-friendly materials. In Proceedings of the 2016 ACM International Symposium on Wearable Computers (ISWC '16). Association for Computing Machinery, New York, NY, USA, 16–23. DOI:10.1145/2971763.2971777

[94] Suen, M.S., Lin, Y.C. and Chen, R., 2016. A flexible multifunctional tactile sensor using interlocked zin nanorod arrays for artificial electronic skin. Procedia Engineering, 168, pp. 1044–1047. DOI: 10.1016/j.proeng.2016.11.336

[95] Kim, H.J., Sim, K., Thukral, A. and Yu, C., 2017. Rubbery electronics and sensors from intrinsically stretchable elastomeric composites of semiconductors and conductors. Science Advances, 3(9), p. e1701114. DOI: 10.1126/sciadv.1701114

[96] Taube Navaraj, W., García Núñez, C., Shakthivel, D., Vinciguerra, V., Labeau, F., Gregory, D.H. and Dahiya, R., 2017. Nanowire FET

based neural element for robotic tactile sensing skin. Frontiers in Neuroscience, 11, pp. 501. DOI: 10.3389/fnins.2017.00501

[97] Byun, J., Lee, Y., Yoon, J., Lee, B., Oh, E., Chung, S., Lee, T., Cho, K.J., Kim, J. and Hong, Y., 2018. Electronic skins for soft, compact, reversible assembly of wirelessly activated fully soft robots. Science Robotics, 3(18), p. eaas9020. DOI: 10.1126/scirobotics.aas9020

[98] Lopes, P.A., Paisana, H., De Almeida, A.T., Majidi, C. and Tavakoli, M., 2018. Hydroprinted electronics: Ultrathin stretchable Ag–In–Ga E-skin for bioelectronics and human–machine interaction. ACS applied materials & interfaces, 10(45), pp. 38760–38768.

[99] Sun, X., Sun, J., Li, T. et al., 2019. Flexible tactile electronic skin sensor with 3D force detection based on porous CNTs/PDMS nanocomposites. Nano-Micro Letters 11, 57. DOI: 10.1007/s40820–019-0288-7

[100] Zolfagharian, A., Kaynak, A. and Kouzani, A., 2020. Closed-loop 4D-printed soft robots. Materials & Design, 188, p. 108411. DOI: 10.1016/j.matdes.2019.108411

[101] Cheng, J., Sundholm, M., Zhou, B., Hirsch, M. and Lukowicz, P., 2016. Smart-surface: Large scale textile pressure sensors arrays for activity recognition. Pervasive and Mobile Computing, 30, pp. 97–112. DOI: 10.1016/j.pmcj.2016.01.007

3

Tactile Sensors for Smart Human–Object Interactions: Devices and Technologies

Andrea Adami* and Leandro Lorenzelli

Fondazione Bruno Kessler, Center for Materials and Microsystems,
Via Sommarive, Trento, Italy
E-mail: andadami@fbk.eu
*Corresponding Author

In the last years, the progresses in microfabrication and printing technologies, in artificial intelligence, and also in basic studies on human tactile perception have opened up new opportunities in the field of tactile sensing, starting from the improvements in dexterous and smart manipulation in robots and moving toward human–object interactions. New perspectives emerged in the monitoring of interactions in several fields including surgery, prosthetics, robot-assisted operations, and smart interaction for augmented reality. In this chapter we briefly introduce the opportunities in this field to discuss the major requirements and challenges for tactile sensors. Then, the major achievements in the field of tactile sensor devices and technology are discussed, including the implementation of the most used tactile sensors' classes and the novel concepts to provide additional features and a full tactile feedback from a sensor, with the final goal of mimicking the surprising capabilities of human skin to sense the surrounding environment.

3.1 Introduction

The research on tactile sensing started from more traditional force and torque sensors in robotics to provide a feedback to robot operations; then, improved tactile sensing emerged as a need to overcome the limitations of initial approaches and moving toward dexterous manipulation in industrial robots.

47

The goal was to move from fixed coding of robot operation to more flexible, smart, adaptive, and fault-tolerant manipulation of objects during the process [1], and safe human–robot interactions [2].

Building on these results, more applications have become possible in different fields. In smart human–object interactions, an almost obvious application was to use robots as an intermediate tool between humans and the operative environment, for instance for intervention in critical or dangerous area [3], or in robot-assisted medicine [4]. In such fields, tactile sensing is the front block of a haptic chain that is able to provide the necessary input to drive the decision and the actions of the operator. Further, this concept can be linked to augmented reality fields, where a tactile system is able to provide additional inputs and capabilities to the operator by multimodal integration [5]. Simulation and training of surgery operations is also a field of application where sensors can be used in a haptic system [6].

In the medical field, miniaturized tactile sensors are often required in force and compliance measurement tasks of the tissue for minimally invasive surgery operations including laparoscopy, thoracoscopy, arthroscopy, ophthalmic microsurgery endoscopy [7, 8]. Moreover, the use of prosthetic limbs equipped with large area of tactile sensors at the socket can be used to select the best configuration, to personalize the prothesis or to evaluate the performance and improve the comfort of the patient [9].

A more challenging progress is to develop a robotic prothesis that is able to provide a feedback to manipulation to improve dexterous interactions. This field needs to cope up with challenges beyond tactile sensing itself, especially in the haptic feedback or in the efficiency and stability of the interface to the neural system to be really effective; after the first proof of concepts the work is progressing with encouraging results toward this frontier of tactile sensing applications [10, 11].

By changing the paradigm, it is also possible to consider the tactile interaction from the object side, i.e. providing tactile sensing to the object. For instance, smart human–machine approaches can be used to monitor and correct gait and posture in advanced sport training and rehabilitation. This is typically achieved not only by the use of accelerometers but also smart devices that are able to map the pressure distribution in the time domain and provide a useful and measurable feedback on the rehabilitation results or a diagnostic tool for technical training for advanced athletes [12, 13].

In the Internet-of-Things (IOT) field, smart objects that are able to sense the interaction between objects and persons can be used to provide additional services or features and adapt their function to the context [14]. Recently, the IoT concept is evolving toward the "tactile internet" concept, where the improvement of communication systems is concerned with low roundtrip latency in conjunction with ultrahigh reliability so as to provide a haptic feedback to the operator of robotic systems equipped with tactile sensing. The goal is to enable remote presence around the globe by the use of smart haptic systems and robots, for instance in the field of assistance for the elderly or telemedicine and for the operation and maintenance of equipment in the industry, especially in critical and dangerous settings [15].

Despite vision systems being more widely exploited in many applications and artificial intelligence applied to image processing could solve many challenges, tactile sensing has many clear benefits and good complementarity to smart cameras, because it can provide more information on the three-dimensional (3D) shape of objects without visual occlusion or lighting issues, and the physical properties of objects are often difficult to be visually evaluated in unknown environments or in untrained settings. On the negative side, tactile acquisition is typically slower than visual because of limited number of sensors and the need to explore sequentially different positions on the object [16].

Therefore, tactile sensing can play a role in many fields of application in stand-alone systems or in combination with visual systems to enable smart interactions. In human–object interactions, tactile sensors can provide much more information than traditional force or torque sensors placed at structure joints, and they can be used with different objectives and with different modalities according to the type of desired interaction, for example, exploring the geometry of an object or an environment, or probing the properties of an object such as temperature, stiffness, roughness, surface properties in general, weight, etc. [1, 16]. Dynamic interaction with objects requires controlling the grasping methodology of the object to avoid slipping or crushing the object and the dynamic forces during the manipulation to drive the action. A different field of application is related to low resolution, large-area sensing to provide collision detection for instance of a robotic arm in an uncontrolled environment. In this case, proximity sensors also can be used without the need for a precise analysis of the contact. On the opposite end of resolution scale, it is possible to consider fingerprint sensors [17] as a niche application of contact mapping, where the resolution needs to be between 50 and 100 µm and pressure is very light (typically 5–100 kPa).

All of these applications may benefit from tactile sensing, if the sensor is designed with the requirements in mind. For this purpose, the selection of the sensing principle, of the technology and design is crucial for successful implementation.

The sensor features need to be adapted to the task requirements. In some cases, single-point sensing or a relatively simple pressure or force mapping approach may be enough to evaluate the contact position and force distribution of the interaction, for instance in the geometrical exploration of objects. Monitoring the dynamic evolution of the contact provides additional insight on object vibration, stability of the grasp, penetration of the gripper into the object, to give a few examples. High resolution mapping may be used to evaluate the roughness and texture of the surface, although the computational effort and hardware required for such large data acquisition and processing are not very cost effective in a broad sense.

Extending the sensing capability beyond perpendicular forces, i.e. from pressure sensing to triaxial force detection at the interaction surface gives more information about the in-plane components of forces for the evaluation of grip, exploration of surface friction coefficient, or the detection of slip of the object.

Since the contact depends on the properties of both the bodies involved, sensing can also include the mapping of stress or deformation in the sensor structure itself. This concept can be used in a bioinspired concept to mimic the skin-sensing capabilities, where the state of stress at each point is monitored by different receptors and used to infer object properties locally; for instance, mapping the local skin stretch can provide an evaluation of the presence of a texture below the spatial resolution of the sensor or a more precise detection of edges [18]; it can also provide information on nonuniformity of stiffness, which for instance is a very important topic in exploration of tissues for medical applications. In this approach, it is possible to evaluate the presence of surface features using lower computational power than required using a high-resolution pressure mapping approach.

The frequency domain response is also very important to detect important information of objects or of the interaction, in addition to the detection of the vibration of the object itself. For instance, the time domain or frequency response can provide information about an impact of the object with the environment. This is a quite peculiar trait of the human species in general, which enables complex interactions with the environment mediated by tools, for instance using tapping exploration to infer the hardness and position of a surface [19] and dexterous use of tools in general. Frequency domain signatures can be also used to detect slipping or texture properties of the surface during surface exploration [20].

Multimodal sensing can also include other parameters such as temperature, electrical and thermal conductivity, and other parameters specific of the application, which can be used in combination with smart data processing to interpret the environment and manage the interactions.

Of course, the collection, management, and processing of data has huge importance in sensitive and efficient tactile multimodal sensing to provide interpretation of data and extraction of the information. This topic is beyond the scope of this chapter, whose focus is to define the selection of sensing principles and technologies more adapted for the required task.

For this purpose, Section 3.2 introduces tactile sensing technologies and devices, comparing different sensing mechanisms, materials, and technologies.

3.2 Technologies and Devices

3.2.1 Fabrication Technologies

In the selection of the best technology for a specific application, the first choice is to define the sensing approach among the options available, i.e. pressure mapping, triaxial contact force sensing or mapping, advanced sensor deformation mapping, and frequency/multimodal feature extraction in the time or frequency domain.

As usual in technology selection, figures of merit and characteristics are defined to select and drive the development of the technology, beyond the obvious general sensor specifications such as spatial resolution, sensitivity, resolution, range, linearity, hysteresis, and stability. The frequency domain feature extraction and dynamic features or high-speed scanning of large arrays may require a sufficient bandwidth of the sensor and acquisition system, and may dictate the choice of sensor concept and materials, since for instance soft substrates or covers act as mechanical filters and may exclude important information in the high-frequency range.

The technology needs to be selected also with manufacturability and integration in mind. In this regard, the selection of materials for the substrate, the functional components of the sensors, and the auxiliary components such as wires, protective coating needs to be evaluated carefully.

The substrate selection is of course related to the choice if the sensor will be rigid or needs to be flexible or stretchable. In the first case, more traditional substrates from microelectronics can be used, for instance silicon or other ceramic piezoelectric substrates like quartz and like lithium niobate

(LiNbO$_3$), to provide devices with high-quality standards of micro- and nanofabrication. Semiconductor properties such as high stress sensitivity, embedded electronic circuits for transduction and amplification, optical conversion in direct SC as III-V, low parasitic effects, low power and high speed can be achieved in standard microfabrication. Stability, reproducibility, and the integration of well-known coatings and materials with stable properties are also advantages of these techniques. The usual drawbacks of these materials are the high stiffness, the brittleness, and the cost, especially for large-area devices.

The complete overview of microfabrication technologies is of course beyond the scope of this chapter and the reader is referred to the many excellent specific books available on the topic, e.g. [21]. In general, the microfabrication technologies are based on layered structures of semiconductor, dielectrics, and metals deposited on wafers with diameter ranging from 25 to 450 mm, where the structures are defined by photolithography and etching processes and in some special case by direct writing, especially for nanometric features in the active area of the sensor. Microelectronics are implemented by the addition of tight control of semiconductor doping by ion implantation, while mechanical structures typical of MEMS devices are based on etching either of the substrate or of structural layers on the wafer. With this approach it is possible to integrate mechanical structures at micron and nanometer scales with sensing units, integrated readout, multiplexing and processing electronics, and functional materials exploiting the technology and equipment originally developed for microelectronics. As an example, in Ref. [22] the integration of piezoelectric layers on top of a silicon transistor (i.e. polymer-oxide-semiconductor FET or "POSFET" structure) can be used to transduce the charge generated by a piezoelectric layer to a more robust signal, and signal conditioning and addressing circuitry can be integrated on chip to provide a more efficient reading.

MEMS technologies are the baseline for what concerns the high resolution, high performance integrated sensors and still have an edge in some applications, but tactile sensing is often related to surfaces of several square centimeters, which is a challenge for the traditional microelectronics and MEMS technologies because of the limited size of the processing area and of the typical high cost per unit area. The cost is a result of the complex equipment and related running cost, and by the high-quality substrate itself. Recently, the production of large area, relatively low resolution (with respect to electronics standards), eventually flexible tactile sensor has progressively moved from lithography-based processes to printing for both active materials and mechanical structures, following the trend of printed electronics, which drove the technological development also in the field of tactile sensors. Integrating a sensor in a "smart hand" or curved object geometry may benefit from a conformable, bendable, or stretchable device that is able to adapt to the surface or from direct sensor printing on the mechanical part, with clear advantage with respect to a more traditional mechanical assembly procedure to embed it in the system. In this perspective, printing and 3D printing may come in handful to provide an advanced integration of sensor into the system, which may be impossible with other methods. Wiring itself can be an issue, especially when low or no computational capability is available on the sensing site. Large area, possibly high-resolution sensors also require to have a strict control on power consumption of the tactile pixel (or "taxel").

The availability of large-format printing techniques including inkjet, screen printing, and in general roll-to-roll facilities moved the focus from the high spatial resolution typical of electronics to material-based approaches, where the functionality rely more on advanced material properties and relatively simple structures. Materials moved from traditional chemical and physical vapor deposition on wafer substrates, which, however, can still be used in large formats, toward solution-based processes on flexible substrates to provide functionality to devices with low cost and highly scalable processes. The advantages of printing are also in the flexibility of design and customization provided by on-demand digital processes such as printing. Of course, flexible sensors are inherently more robust, besides the possibility to adapt to a curved object, due to the absence of brittle materials in the structure. The novel, flexible electronic technologies enable the implementation of sensor and devices by the use of polymeric substrates and functional materials [23], including organic semiconductors [24], semiconducting oxides such as indium–gallium oxide (IGO) and indium–gallium–zinc oxide (IGZO) [25, 26], graphene and piezoelectric polymers such as poly-vinylidene difluoride, and its copolymers (e.g. the mostly used copolymer with tri-fluoroethylene, PVDF-TrFE). Flexible electronics are typically slower because of lower carrier mobility in flexible semiconductors, especially organic ones, although there was a good progress in performances in the last years. Low stiffness can result in mechanical filtering of high-frequency components. Another limit is the need for low-temperature processing compatible with the substrates, which is a limit to the material processing options and the stability in operative conditions. Among the mostly used substrates, polyimides can reach 400°C while other polymers (e.g. PET, PVC) have lower thermal resistance.

In any implementation of a tactile sensor, the sensor robustness to mechanical contact, temperature, and environment in general is a strict requirement since the sensor by definition needs to come in contact with bodies and potentially harsh conditions, and the options for increasing the protection may be limited. For instance, coating thickness and stiffness is strictly related to the stress mapping resolution and sensitivity that can be achieved by the sensor; therefore, the chemical and mechanical resistance of the sensor is one term of the trade-off of performance vs. robustness and needs to be carefully evaluated in the specific case.

3.2.2 Tactile Sensor Devices

3.2.2.1 Piezoresistive and resistive MEMS

Pressure or strain sensitivity is intrinsic to several structures, including metallic wires, silicon, and nanostructured materials, and can be used to fabricate tactile sensors. To compare the performance of different materials, it is convenient to use the gauge factor (GF), which is defined as

$$GF = (\Delta R/R)/\varepsilon \tag{3.1}$$

The change in resistance is the result of the geometrical deformation of the resistor and the variation of the resistivity of the material. The relative variation of resistance induced by a strain ε could be calculated as

$$\Delta R/R = (1 + 2\nu)\varepsilon + \Delta\rho/\rho \tag{3.2}$$

where ν is the Poisson's ratio and $\Delta\rho/\rho$ is the relative variation in the resistivity. The geometric effects depend only on the Poisson's ratio and provide a typical GF between 1.6 and 1.9. The variation of the resistivity instead is material dependent and the overall GF usually is in the order of 2–5 for most metallic strain gauges. Typical temperature coefficient, TCR, is in the order of 1000–4000 ppm/K for metals, with the exceptions of low TCR alloys such as constantan. Different gauge grades are provided to match the thermal expansion of the substrate with the thermal change of resistance of the resistor, by selecting changes in alloy composition and treatment.

The GF in a single crystal silicon strain gauges reaches values up to 100, due to the predominant piezoresistive change of material resistivity, which depends on the crystal structure of the material, the direction of the strain, and on the doping concentration [27, 28]. In particular, p-type silicon has good response to strain in [110] direction and n-type in [100] direction, while for both the sensitivity in direction rotated by 45 degrees is negligible; further,

Table 3.1   Silicon properties for low doping

Materials	n-Si (10^{-11} Pa^{-1})	p-Si (10^{-11} Pa^{-1})
π_{11}	-102.2	6.6
π_{12}	53.4	-1.1
π_{44}	-13.6	138.1

shear stress can provide high response, depending on the doping and the orientation. The response of a piezoresistor to an arbitrary stress state and orientation of the sensing element needs to be calculated from the piezoresistive coefficients represented in a tensor notation. The change of resistivity for a current along direction (ω) and stress along direction (λ) [27] is:

$$\frac{\Delta \rho_\omega}{\rho} = \sum_{\lambda=1}^{6} \pi_{\omega\lambda} \sigma_\lambda \tag{3.3}$$

The properties also depend on doping level, where the higher sensitivity to stress and temperature is achieved at low doping (e.g. 10^{15} at/cm^3).

It is possible to find an optimal doping to increase the stress to temperature sensitivity, which is typically around 10^{19} at/cm^3. The response is completely linear up to 0.3% and it has been successfully measured for strain up to 1%, although the brittle nature of silicon and the eventual presence of defects can result in failures if the device is not properly designed and handled.

By reducing the thickness of the silicon substrate, it is possible to reduce the overall stiffness of the substrate to be compatible with an arbitrary radius of curvature [29]. While the bending stiffness scales with the third power of the thickness, the stress at chip surface scales with the thickness at a given curvature. Nevertheless, it is worth mentioning that the thin silicon is still brittle by nature, i.e. subject to catastrophic failure if the material strength or the fracture toughness is reached, especially for point- loads or contacts with sharp and hard objects. Another limitation is that the thin silicon is bendable but not stretchable or conformable. Therefore, the thin silicon concept is particularly useful to integrate in flexible structures the advantages of single crystal silicon, especially the carrier mobility to provide high speed and miniaturized electronics, with reduced impact to system flexibility and thickness. This is therefore especially useful for addressing taxels in large sensor arrays and data processing on site. It can be also used for signal transduction, but the trend is to use flexible materials directly, even if they cannot provide the carrier mobility of silicon.

In polysilicon, the random orientation of the crystals and the presence of grain boundaries reduce the GF to values in the order of 20–40 in all directions. However, the performances are better than metallic strain gauges and the cost of production lower with respect to standard crystalline silicon. In addition, in the case of amorphous and microcrystalline silicon the layer can be also deposited at low temperature on flexible substrates with gauge factors in the order of 20–30 [30–32].

A GF higher than the geometrical GF can be also achieved by conductive nanoparticles or nanotube dispersed in a polymeric matrix. The properties of these materials are described by a percolation law, where the percolation threshold is used to describe the critical concentration above which the material becomes conductive. The particles form conductive paths in the material, whose resistivity is the statistical average of particle conductivity and tunneling resistance across the gap between particles. When the material is deformed, the gaps experiment an exponential change of resistance as a function of the width variation, as it is expected from tunneling. This behavior is confirmed by the fact that piezoresistivity is usually higher for lower particle concentrations, where resistivity is dominated by the contact between particles or wires. Gauge factors of 20 can be achieved in best cases, while the high concentration sensitivity is similar to gauges made of metals or conductive materials in general. In these materials the production process parameters including stirring, priming and polymerization are critical to the achievement of target resistivity and properties in general [33, 34]. The use of these materials is especially useful for printed, flexible devices, where the polymeric matrix (e.g. PDMS) is fully compatible with the substrate properties.

Other nanostructured materials can provide gauge factors beyond the plain geometrical response by exploiting tunneling conduction at grain boundary similar to the case of filled polymers. One class of such materials with piezoresistive properties is the cermets, which are composite materials formed by a metallic phase and a ceramic phase, for example Au or Pt in SiO_2. If the metallic phase is formed by particles dispersed in the ceramic matrix, the GF can reach values around 20. In the opposite case the material is a conductor with resistivity modulated by the phase proportion [35, 36].

There are different strategies to build a tactile sensor with stress-sensitive resistors. The direct sensing of pressure on the resistor surface, i.e. perpendicular to the current flow is typically lower than longitudinal sensing, but it is still possible with silicon due to the higher coefficient and 3D distribution of the response with respect to crystal planes, while for metallic strain gauges the response is too low. In general, the performances of direct pressure sensing are not very good and usually a 3D structure is used to increase the sensitivity, by converting the forces at the surface into high longitudinal strain of the sensing element, for instance by the use of cantilevers, beams, or membranes with the sensing elements placed in the high-stress region, typically at the anchoring points. The structures can be macroscopic for low-resolution sensors or MEMS for high resolution, integrated sensors [37].

More complex arrangements can provide other features such as triaxial sensing capability by the use of an off-plane structure converting a tangential force into a moment and a differential response on the elements reacting to the structure tilt. This configuration can use suspended beams or membranes to support the central unit receiving the contact force [38].

A similar approach can be implemented using an array of polymeric domes or bumps; however, the reaction of the structure to a tangential force results in a pressure distribution between the domes and the actual sensor surface. In this case, four corner elements and eventually an additional central sensitive element (e.g. membrane-based pressure sensors) can be used to process the triaxial force information. In such cases the resolution is lower because of the multielement taxel configuration but microfabricated sensors are still compatible with standard requirements (typically 1 mm in high-resolution applications) [39].

3.2.2.2 Capacitive

Capacitive sensors are one of the simplest configurations of pressure sensor, and are typically implemented by the use of two electrodes and a compressible dielectric layer or a membrane and an air gap. The sensitivity of these sensors is defined as S = 1/C dC/dt and is proportional to −1/t where C is the capacitance and t is the electrode spacing. Since the change of thickness is dt = ϵt, the response is defined by the strain ϵ of the capacitor under the load. It is possible to demonstrate that if t << L with L the size of the taxel, the lateral deformation is limited by the structure and the Poisson ratio can change substantially the apparent stiffness of the material, especially for Poisson approaching 0.5 as in the case of elastomers. In this case, the overall sensitivity is (P is pressure, E is Young's modulus, and ν is Poisson's ratio)

$$\frac{\Delta C}{C} = \frac{P}{E}\left(1 - \frac{2\nu^2}{1-\upsilon}\right) \tag{3.4}$$

For this reason, a low E material is needed to achieve a good sensitivity, e.g. an elastomer such as silicone rubber like PDMS, in which E can range

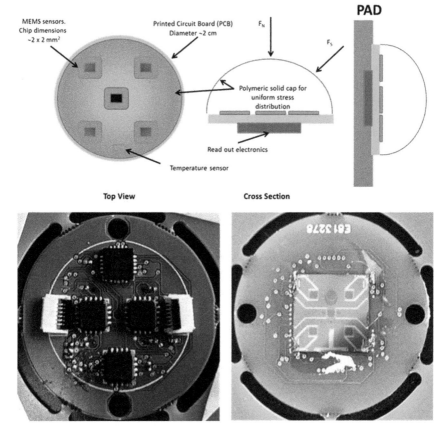

Figure 3.1   MEMS triaxial force sensor, adapted from Refs. [9, 39]. (a) A sensor concept and (b) Packaging of the silicon sensor prototype.

from 0.6 to 3.6 MPa. However, elastomeric foam or an array of grooves or bumps can be used to further reduce the apparent E and to avoid the negative effect of lateral confinement [37].

The proof of fabrication principle of these sensors is typically confirmed by casting or molding of polydimethylsiloxane (PDMS), but for large area and mass production printing and roll-to-roll techniques may be used. Typically, it involves the deposition of metal contact by PVD (e.g. evaporation or sputtering) or printing (e.g. screen or inkjet printing) and lamination of the different layers, which may be eventually patterned [41]. The cost of this technology can be extremely low, but the drawbacks are the resistance of the structure, especially the integrity of metallic wires on top of a soft layer, the

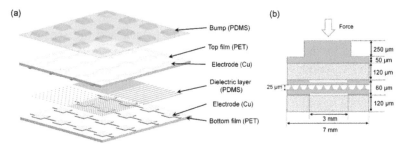

Figure 3.2 Patterned PDMS dielectric for high sensitivity capacitive sensors. Reproduced from Ref. [40]. (a) Exploded view of the capacitive tactile sensing array. (b) Cross-sectional view of one sensing unit, and (c) Schematic view of different geometries of the microstructures on the polydimethylsiloxane (PDMS) layer, including pyramids with spaces of 50 μm (type I) and 150 μm (type II), and V-shape grooves with spaces of 50 μm (type III) and 150 μm (type IV).

nonlinear mechanical properties of rubbers or patterned layers, the hysteresis and speed of mechanical response of polymers, foams, and patterned rubbers. Contacts can be also provided by stretchable interconnections such as conductive polymers and nanowire composites to increase the sensor reliability [41, 42]. In addition to flexibility, the advantage of these techniques is the possibility to work on large area sensor arrays and the easier customization of the device for specific shapes of the area to be equipped with sensors.

Another approach is the use of a membrane carrying one electrode and a second electrode across an air gap. In this case the sensitivity can be adjusted with the membrane size and thickness, as well as the electrode spacing, but the structure becomes more complex to fabricate. The design needs to use the plate theory for analytical calculation or an FEM model. For instance, the deflection w for a circular plate with clamped edges [43] is

$$w\left(r\right) = -\frac{P}{64\,D}\left(a^2 - r^2\right)^2 \tag{3.5}$$

$$D = \frac{E\,t^3}{12\left(1 - \nu^2\right)} \tag{3.6}$$

where t is the membrane thickness, P is the pressure, a is the membrane radius, r is the radial position, and D is the flexural stiffness of the membrane. The sensor can be designed to optimize response, pressure range, and maximum load to failure.

The implementation of membranes by bulk or surface micromachining can be easily achieved by anisotropic wet etching or deep- Reactive Ion

Etching (RIE) in the silicon wafer thickness. Alternatively, the membrane can be implemented with a deposited thin film stack and the sensor gap can be provided by the etching of a sacrificial layer beneath the membrane structure. The electrodes can be provided by doping the silicon itself or by deposition of a metal layer; wafer to wafer bonding techniques can be useful to provide metallic electrodes on both sides of the gap. The fabrication of suspended plates with silicon-on-insulator wafers provides the advantages of single crystal silicon with the most accurate thickness control, but the cost of substrate is typically one order of magnitude higher [44].

Capacitive membranes can be also implemented by lamination, patterning, and etching of polymers to provide the mechanical structure, and using deposited metal layers as an etch stop and for the sensing units [45]. This process can also integrate a multimodal sensor including piezoresistive elements.

3.2.2.3 Piezoelectric

Piezoelectric devices are widely used for tactile sensing. Piezoelectric materials are materials where the deformation and the polarization of the material and Hooke's law are linked by the following Equations [46]:

$$D_i = d_{ij}T_{jk} + \varepsilon_{ij}^T E_j \tag{3.7}$$

$$S_{ij} = s_{ijkl}^E T_{kl} + d_{kij}E_k \tag{3.8}$$

where D is the electric displacement, d_{ij} are the piezoelectric coefficients in the strain-charge form, T is the stress, S is the strain, s^E is the mechanical compliance, ϵ is the permittivity, and E is the electric field. It is possible to use alternative combinations of the physical quantities involved; for example, the e coefficients are used to relate stress to the applied electric field. Depending on the crystal symmetry planes, only few coefficients are relevant; typically, for the commonly used materials with the exception of quartz, the coefficients d31, d33, and d15 are the most important. When it is used as an actuator, a vertical polarization is used to generate a strain in the vertical direction in the 33 mode or in the horizontal direction in the 31 mode. For sensing applications, the vertical or planar deformation of the material can result in a vertical polarization of the piezoelectric element coupled to a mechanical structure as described for piezoresistive sensors.

The most common materials for these sensors are ceramic high performances materials like PZT or lithium niobate, or polymeric, where PZT-TrFE copolymer blends represent the vast majority of literature cases due to its

Table 3.2 Examples of piezoelectric properties for comparison

Materials	d31, pC/N	d33, pC/N	ϵ_{rel}	E, GPa	T curie, °C
PZT-5H	−274	593	2000	64	350
Lithium niobate	−3	9.3	85	200	1000
PMN-PT	−600	1600	1500	120	150
AlN	−3	5	9	350	2000
AlScN (40% Sc)	−13	30	16	230	2000
PVDF-TRFE	15	−20	10	2	100

easier polarizability with respect to unblended PVDF and its relatively high piezoelectric coefficients. AlN and its composites with scandium (AlScN) are gaining much interest due to their relatively good performances for integrated sensing applications, especially with the addition of scandium up to 40%, and their compatibility with complementary metal–oxide–semiconductor (CMOS) processes [46, 47]. At Curie's temperature the piezoelectric effect disappears, but the maximum operative temperature is usually lower. The material properties heavily depend on composition and processing especially for polycrystalline thin films, therefore it is possible to find very different values in the literature. Here, for comparison only, we report a table with typical properties of most used materials (Table 3.2).

Ceramic materials are usually processed at very high temperatures, typically above 600°C for sputtering, or even higher if the layer needs to be sintered after deposition as in the case of printed layers, where the sintering temperature can be commonly above 900°C [46]. Polarization process by high electric field or corona poling are needed to orient the piezoelectric domains and provide polycrystalline materials with the required properties.

The advantage of piezoelectric tactile sensor is the possibility to have a direct self-powered charge generation in response to an applied stress, with a relatively simple capacitor structure, which can be easily read with an electronic circuit. The response of piezoelectric tactile sensors is particularly sensible to force variations, since the charge is generated during the force application; however, in static conditions the charge on the capacitor can be dissipated by parasitic losses, especially with PVDF-TrFE. For this reason, the piezoelectric devices are typically used in frequency bandwidth above 1 Hz [48] and are not suitable for static force detection. Frequency-based approaches such as vibration, slip and impact detection, and texture recognition are very much suited for piezoelectric sensors. Ceramics may allow a wider bandwidth at both ends of the range because of lower parasitic conduction and higher mechanical stiffness.

A different approach is to use the piezoelectric actuator to probe the surface by dynamic methods. The approach was originally developed for macroscopic probes, typically cantilevers, in medical applications such as tissue stiffness probing, which is related to the identification of tumors from normal tissue [49] or to evaluate the myocardial functionality [50]. The concept is based on the change of mechanical behavior of an actuator when it is in contact with surfaces with different mechanical properties, which can be modeled by an equivalent damping, elasticity, and mass of the contact surface [51–54]. In brief, the resonance frequency of the actuator changes with the external load and surface stiffness, while the damping of the surface results in an attenuation of the signal.

3.2.2.4 Other sensing techniques

Magnetic fields can be used to implement a tactile sensor. In principle, the measurement is based on a magnetic field source, which can be a coil [55], a permanent magnet [56], or a nanocomposite layer of structures e.g. a pillar forest [57], and a magnetic field sensor implemented in the substrate. The sensing elements can be implemented for instance by using coils [58], Hall sensors [56], or giant magnetoresistive sensors [55, 59]; however, from the change in magnetic field distribution it is possible to infer the displacement of the magnetic element in the top deformable layer due to the contact with an object. While Hall and giant magnetoresistive sensors measure the magnetic field intensity, pickup coils detect only the field variations, thus providing a derivative type of a tactile sensor.

Optical tactile sensors can be implemented with different approaches: one of the most common concepts is to use a camera to evaluate the deflection of an array of features at the contact, where the mechanical part is passive and the sensing is implemented with vision-based pattern recognition and classification algorithms [60]. A second option is to use fiber-based strain gauges, typically by using a Bragg grating implemented on the fiber cladding: the deformation of the grating due to fiber stretching induces a shift of the reflected wavelength [61]. By selecting different wavelengths, it is possible to read a number of elements on a single fiber, then providing a large array of taxels. Transmission properties of fiber, for instance the losses of a fiber as a function of the bending radius, can be also used to evaluate the deformation of a pixel unit [62]. The immunity to electromagnetic interferences is one of major advantages of optical techniques.

A few fluidic concepts can be also found. For instance the compression of channels into an elastomeric layer can be detected by the change of resistance of the fluid network filled with an appropriate fluid [63]. The advantage is the high intrinsic reliability of the sensing element due to its virtually infinite deformability.

3.2.2.5 Recent trends

As discussed earlier, there are more opportunities for tactile sensing beyond the pressure mapping at the contact. Probing the properties of an object may also include the evaluation of the mechanical properties of the surface. In addition to vibrational piezoelectric probing cited earlier, it is also possible to evaluate the hardness of an object by the use of differential reading on pixel with different response when put in contact to a surface with different properties. Conceptually, if the response of the pixel depends on its deformation, and its stiffness is similar to the surface under analysis, then both the deformation of the external body and sensor unit are proportional to their relative compliance. If the sensor is composed by taxels with different stiffness, the differential readout can provide information about the touched surface. For instance, a soft body tends to distribute a load more uniformly on structures with alternated stiff and compliant areas, thus providing an extra response on the membrane [64] or the compliant areas of coating [65] with respect to the rigid reference elements. Differential MEMS spring [66, 67] or membranes with different size or stiffness [68] connected to pillars may be used to detect the degree of indentation into the external body. Active indentation can be also provided by magnetic or pneumatic actuation that in turn is able to provide curves similar to classical load-displacement testing of materials [69, 70]. The force reading can be provided with piezoresistive, piezoelectric, or capacitive sensing units as described earlier.

Slip detection is an intrinsically dynamic measure suited for piezoelectric, triaxial resistive tactile sensors or in general if the bandwidth of the sensor and of the readout electronics is sufficient. The contact force is a triaxial load, where the slip happens when the tangential force exits from the "friction cone" i.e. exceeding the maximum static friction force. Since the dynamic friction coefficient is typically lower than the static one, the detection of incipient slip is based on the microvibrations due to local and intermittent movements of the object on the sensor followed by sticking steps, which result in a specific spectrum of noise in the force measurement, especially in the tangential direction, but axial components can be also used. The evaluation of the signal spectrum is specific of the contact surfaces and

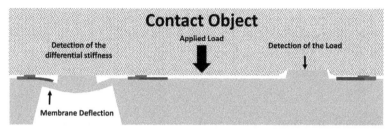

Figure 3.3 Differential stiffness structures for tissue stiffness measurements similar to the concept in Ref. [45]. Hardness sensor schematic (a) cross section, with membrane and bulk hardness sensors and (b) in contact with an object, the sensors deform, with apparent pressures proportional to the contact object hardness.

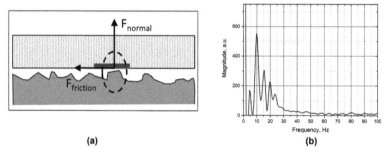

Figure 3.4 Frequency-domain sensor for slip and surface pattern recognition: (a) Scheme of forces at sensor interface, where friction force is defined by roughness, deformability of sensor and surface, materials, and speed of slip; (b) conceptual FFT of sensor output, showing a fingerprint specific of the slipping interaction between the sensor and the surface.

an appropriate pattern recognition algorithm can be used to infer both the incipient slip and surface properties such as roughness, presence of patterns, and friction coefficient [71, 72]; the patterning of the sensor surface modifies the contact properties and can be used to optimize the detection process or extract more information from pattern recognition algorithms [73].

In addition to tactile mechanical sensing, it is also possible to add other features to the sensor, for instance the evaluation of temperature and of thermal conductivity and specific heat by the use of appropriate combination of resistive heaters and thermometers [45]. This concept can be useful to provide a perception of temperature more consistent with human sense of touch, since thermal conductivity is important in the human perception of heat. The integration of multimodal sensing including pressure, 3D forces, surface hardness and curvature as discussed earlier can therefore provide full tactile information.

3.3 Conclusions

Tactile sensing showed an impressive progress since the appearance of the first concepts for robotic applications using mechanical assemblies to measure the forces involved in object manipulation. With the help of miniaturization and integration provided by the MEMS technology, the field has moved to high resolution and high precision pressure mapping by the implementation of several sensing elements concepts and tactile sensors functionalities. The next step was moving toward a more bioinspired concept of tactile feedback including for instance the analysis of grasp stability by the use of slip detection, of surface properties such as stiffness, texture and friction. Then, technologies related to large-area electronics and digital manufacturing such as roll-to-roll and 3D printing have further stimulated the field enabling more opportunities for large area, low cost electronic skin applications and industrial case studies. Now, the field is ready to take on challenges related to real applications and the interface to biological system, with the grand objective to provide effective, natural, and full tactile feedback experience in a number of human–object interaction, medicine and prosthetics applications.

Acknowledgements

This work has been developed in the framework of the H2020-MSCA-ITN-2019 project INTUITIVE (INnovative Network for Training in ToUch InteracTIVE Interfaces). This project is funded by the EU Horizon 2020 research and innovation program under the grant agreement No 861166.

References

[1] Z. Kappassov, J. A. Corrales, and V. Perdereau, "Tactile sensing in dexterous robot hands – Review," *Rob. Auton. Syst.*, vol. 74, pp. 195–220, 2015, doi: 10.1016/j.robot.2015.07.015.

[2] B. D. Argall and A. G. Billard, "A survey of tactile human-robot interactions," *Rob. Auton. Syst.*, vol. 58, no. 10, pp. 1159–1176, 2010, doi: 10.1016/j.robot.2010.07.002.

[3] J. M. Gandarias, J. M. Gómez-de-Gabriel, and A. J. García-Cerezo, "Enhancing perception with tactile object recognition in adaptive grippers for human-robot interaction," *Sensors (Switzerland)*, vol. 18, no. 3, 2018, doi: 10.3390/s18030692.

[4] M. I. Tiwana, S. J. Redmond, and N. H. Lovell, "A review of tactile sensing technologies with applications in biomedical engineering," *Sensors Actuators, A Phys.*, vol. 179, pp. 17–31, 2012, doi: 10.1016/j.sna.2012.02.051.

[5] F. Garzotto and M. Gelsomini, "Magic room: A smart space for children with neurodevelopmental disorder," *IEEE Pervasive Comput.*, vol. 17, no. 1, pp. 38–48, 2018, doi: 10.1109/MPRV.2018.011591060.

[6] P. Saccomandi, E. Schena, C. M. Oddo, L. Zollo, S. Silvestri, and E. Guglielmelli, "Microfabricated tactile sensors for biomedical applications: A review," *Biosensors*, vol. 4, no. 4, pp. 422–448, 2014, doi: 10.3390/bios4040422.

[7] I. K. Takashima and K. Yoshinaka, "Complex Medical Engineering," F. H. Wu J.L., Ito K., Tobimatsu S., Nishida T., Ed. Springer, Tokyo, 2007.

[8] R. Sedaghati, J. Dargahi, and H. Singh, "Design and modeling of an endoscopic piezoelectric tactile sensor," *Int. J. Solids Struct.*, vol. 42, no. 21–22, pp. 5872–5886, 2005, doi: 10.1016/j.ijsolstr.2005.03.029.

[9] W. Xu et al., "Surface Pressure Reconstruction for a Prosthetic Socket Design System – A Numerical Case Study," *IEEE Int. Conf. Control Autom. ICCA*, vol. 2018-June, no. 645239, pp. 941–946, 2018, doi: 10.1109/ICCA.2018.8444193.

[10] G. S. Dhillon and K. W. Horch, "Direct neural sensory feedback and control of a prosthetic arm," *IEEE Trans. Neural Syst. Rehabil. Eng.*, vol. 13, no. 4, pp. 468–472, 2005, doi: 10.1109/TNSRE.2005.856072.

[11] G. Valle et al., "Biomimetic intraneural sensory feedback enhances sensation naturalness, tactile sensitivity, and manual dexterity in a bidirectional prosthesis," *Neuron*, vol. 100, no. 1, pp. 37–45.e7, 2018, doi: 10.1016/j.neuron.2018.08.033.

[12] A. H. Abdul Razak, A. Zayegh, R. K. Begg, and Y. Wahab, "Foot plantar pressure measurement system: A review," *Sensors (Switzerland)*, vol. 12, no. 7, pp. 9884–9912, 2012, doi: 10.3390/s120709884.

[13] P. S. Girão, P. M. P. Ramos, O. Postolache, and J. Miguel Dias Pereira, "Tactile sensors for robotic applications," *Meas. J. Int. Meas. Confed.*, vol. 46, no. 3, pp. 1257–1271, 2013, doi: 10.1016/j.measurement.2012.11.015.

[14] H. Amroun and M. Ammi, "Who used my smart object? A flexible approach for the recognition of users," *IEEE Access*, vol. 6, pp. 7112–7122, 2017, doi: 10.1109/ACCESS.2017.2776098.

[15] M. Maier, M. Chowdhury, B. P. Rimal, and D. P. Van, "The tactile internet: Vision, recent progress, and open challenges,"

IEEE Commun. Mag., vol. 54, no. 5, pp. 138–145, 2016, doi: 10.1109/MCOM.2016.7470948.

[16] S. Luo, J. Bimbo, R. Dahiya, and H. Liu, "Robotic tactile perception of object properties: A review," *Mechatronics*, vol. 48, no. November, pp. 54–67, 2017, doi: 10.1016/j.mechatronics.2017.11.002.

[17] S. Memon, M. Sepasian, and W. Balachandran, "Review of finger print sensing technologies," *IEEE INMIC 2008 12th IEEE Int. Multitopic Conf. – Conf. Proc.*, pp. 226–231, 2008, doi: 10.1109/INMIC.2008.4777740.

[18] V. Hayward, "Is there a 'plenhaptic' function?," *Philos. Trans. R. Soc. B Biol. Sci.*, vol. 366, no. 1581, pp. 3115–3122, 2011, doi: 10.1098/rstb.2011.0150.

[19] M. Strese, C. Schuwerk, A. Iepure, and E. Steinbach, "Multimodal feature-based surface material classification," *IEEE Trans. Haptics*, vol. 10, no. 2, pp. 226–239, 2017, doi: 10.1109/TOH.2016.2625787.

[20] Y. Xin et al., "PVDF tactile sensors for detecting contact force and slip: A review," *Ferroelectrics*, vol. 504, no. 1, pp. 31–45, 2016, doi: 10.1080/00150193.2016.1238723.

[21] M. J. Madou, *Fundamentals of Microfabrication and Nanotechnology, Three-Volume Set*. CRC Press, 2018.

[22] R. S. Dahiya et al., "Towards tactile sensing system on chip for robotic applications," *IEEE Sens. J.*, vol. 11, no. 12, 2011, doi: 10.1109/JSEN.2011.2159835.

[23] S. Khan, L. Lorenzelli, and R. S. Dahiya, "Technologies for printing sensors and electronics over large flexible substrates: A review," *IEEE Sens. J.*, vol. 15, no. 6, pp. 3164–3185, 2015, doi: 10.1109/JSEN.2014.2375203.

[24] Y. Qian et al., "Stretchable organic semiconductor devices," *Adv. Mater.*, vol. 28, no. 42, pp. 9243–9265, 2016, doi: 10.1002/adma.201601278.

[25] Y. Sun and J. A. Rogers, "Inorganic semiconductors for flexible electronics," *Adv. Mater.*, vol. 19, no. 15, pp. 1897–1916, 2007, doi: 10.1002/adma.200602223.

[26] L. Petti et al., "Metal oxide semiconductor thin-film transistors for flexible electronics," *Appl. Phys. Rev.*, vol. 3, no. 2, 2016, doi: 10.1063/1.4953034.

[27] Y. Kanda, "A graphical representation of the piezoresistance coefficients in silicon," *IEEE Trans. Electron Devices*, vol. 29, no. 1, pp. 64–70, 1982, doi: 10.1109/T-ED.1982.20659.

[28] A. A. Barlian, W. T. Park, J. R. Mallon, A. J. Rastegar, and B. L. Pruitt, "Review: Semiconductor piezoresistance for microsystems," *Proc. IEEE*, vol. 97, no. 3, pp. 513–552, 2009, doi: 10.1109/JPROC.2009.2013612.

[29] R. S. Dahiya and S. Gennaro, "Bendable ultra-thin chips on flexible foils," *IEEE Sens. J.*, vol. 13, no. 10, pp. 4030–4037, 2013, doi: 10.1109/JSEN.2013.2269028.

[30] L. Maiolo et al., "Flexible sensing systems based on polysilicon thin film transistors technology," *Sensors Actuators, B Chem.*, vol. 179, pp. 114–124, 2013, doi: 10.1016/j.snb.2012.10.093.

[31] H. C. Lim et al., "Flexible membrane pressure sensor," *Sensors Actuators, A Phys.*, vol. 119, no. 2, pp. 332–335, 2005, doi: 10.1016/j.sna.2004.10.012.

[32] X. Yu, J. Thaysen, O. Hansen, and A. Boisen, "Optimization of sensitivity and noise in piezoresistive cantilevers," *J. Appl. Phys.*, vol. 92, no. 10, pp. 6296–6301, 2002, doi: 10.1063/1.1493660.

[33] Alamusi, N. Hu, H. Fukunaga, S. Atobe, Y. Liu, and J. Li, "Piezoresistive strain sensors made from carbon nanotubes based polymer nanocomposites," *Sensors*, vol. 11, no. 11, pp. 10691–10723, 2011, doi: 10.3390/s111110691.

[34] W. Bauhofer and J. Z. Kovacs, "A review and analysis of electrical percolation in carbon nanotube polymer composites," *Compos. Sci. Technol.*, vol. 69, no. 10, pp. 1486–1498, 2009, doi: 10.1016/j.compscitech.2008.06.018.

[35] C. A. Neugebauer, "Resistivity of cermet films containing oxides of silicon," *Thin Solid Films*, vol. 6, no. 6, pp. 443–447, 1970, doi: 10.1016/0040-6090(70)90005-2.

[36] H. Schmid-Engel, S. Uhlig, U. Werner, and G. Schultes, "Strain sensitive Pt-SiO2 nano-cermet thin films for high temperature pressure and force sensors," *Sensors Actuators, A Phys.*, vol. 206, pp. 17–21, 2014, doi: 10.1016/j.sna.2013.11.021.

[37] C. Chi, X. Sun, N. Xue, T. Li, and C. Liu, "Recent progress in technologies for tactile sensors," *Sensors (Switzerland)*, vol. 18, no. 4, 2018, doi: 10.3390/s18040948.

[38] L. Beccai et al., "Development and experimental analysis of a soft compliant tactile microsensor for anthropomorphic artificial hand," *IEEE/ASME Trans. Mechatronics*, vol. 13, no. 2, pp. 158–168, 2008, doi: 10.1109/TMECH.2008.918483.

[39] L. Lorenzelli, G. Sordo, A. Bagolini, G. Resta, "Socketmaster: Integrated sensors system for the optimised design of prosthetic socket for above knee amputees," *Proc. – 2017 1st New Gener. CAS, NGCAS 2017*, no. 645239, pp. 233–236, 2017, doi: 10.1109/NGCAS.2017.70.

[40] Z. Ji et al., "The design and characterization of a flexible tactile sensing array for robot skin," *Sensors (Switzerland)*, vol. 16, no. 12, 2016, doi: 10.3390/s16122001.

[41] T. Yang, D. Xie, Z. Li, and H. Zhu, "Recent advances in wearable tactile sensors: Materials, sensing mechanisms, and device performance," *Mater. Sci. Eng. R Reports*, vol. 115, pp. 1–37, 2017, doi: 10.1016/j.mser.2017.02.001.

[42] A. Salim and S. Lim, "Review of recent inkjet-printed capacitive tactile sensors," *Sensors (Switzerland)*, vol. 17, no. 11, 2017, doi: 10.3390/s17112593.

[43] S. Timoshenko, S. and Woinowsky-Krieger, *Theory of Plates and Shells*. McGraw-Hill, 1959.

[44] Y. Y. Choi et al., "Enhancement of local piezoresponse in polymer ferroelectrics via nanoscale control of microstructure," *ACS Nano*, vol. 9, no. 2, pp. 1809–1819, 2015, doi: 10.1021/nn5067232.

[45] J. Engel, J. Chen, Z. Fan, and C. Liu, "Polymer micromachined multimodal tactile sensors," *Sensors Actuators, A Phys.*, vol. 117, no. 1, pp. 50–61, 2005, doi: 10.1016/j.sna.2004.05.037.

[46] S. J. Rupitsch, *Piezoelectric Sensors and Actuators*. Springer-Verlag Berlin Heidelberg, 2019.

[47] M. Akiyama, K. Umeda, A. Honda, and T. Nagase, "Influence of scandium concentration on power generation figure of merit of scandium aluminum nitride thin films," *Appl. Phys. Lett.*, vol. 102, no. 2, pp. 0–4, 2013, doi: 10.1063/1.4788728.

[48] R. S. Dahiya, G. Metta, M. Valle, A. Adami, and L. Lorenzelli, "Piezoelectric oxide semiconductor field effect transistor touch sensing devices," *Appl. Phys. Lett.*, vol. 95, no. 3, pp. 3–6, 2009, doi: 10.1063/1.3184579.

[49] F. Ju et al., "A Variable-Impedance Tactile Sensor with Online Performance Tuning for Tissue Hardness Palpation in Robot-Assisted Minimally Invasive Surgery," *Proc. Annu. Int. Conf. IEEE Eng. Med. Biol. Soc. EMBS*, vol. 2018-July, pp. 2142–2145, 2018, doi: 10.1109/EMBC.2018.8512717.

[50] K. Miyaji, S. Sugiura, S. Omata, Y. Kaneko, T. Ohtsuka, and S. Takamoto, "Myocardial tactile stiffness: A variable of regional myocardial function," *J. Am. Coll. Cardiol.*, vol. 31, no. 5, pp. 1165–1173, 1998, doi: 10.1016/S0735-1097(98)00063-1.

[51] S. Omata, Y. Murayama, and C. E. Constantinou, "Real time robotic tactile sensor system for the determination of the physical properties of biomaterials," *Sensors Actuators, A Phys.*, vol. 112, no. 2–3, pp. 278–285, 2004, doi: 10.1016/j.sna.2004.01.038.

[52] Y. Murayama, C. E. Constantinou, and S. Omata, "Development of tactile mapping system for the stiffness characterization of tissue slice using novel tactile sensing technology," *Sensors Actuators, A Phys.*, vol. 120, no. 2, pp. 543–549, 2005, doi: 10.1016/j.sna.2004.12.027.

[53] G. M. Krishna and K. Rajanna, "Tactile sensor based on piezoelectric resonance," *IEEE Sens. J.*, vol. 4, no. 5, pp. 691–697, 2004, doi: 10.1109/JSEN.2004.833505.

[54] P. Brunetto, L. Fortuna, P. Giannone, S. Graziani, and F. Pagano, "A resonant vibrating tactile probe for biomedical applications based on IPMC," *IEEE Trans. Instrum. Meas.*, vol. 59, no. 5, pp. 1453–1462, 2010, doi: 10.1109/TIM.2009.2038297.

[55] M. Goka, H. Nakamoto, and S. Takenawa, "A magnetic type tactile sensor by GMR elements and inductors," *IEEE/RSJ 2010 Int. Conf. Intell. Robot. Syst. IROS 2010 – Conf. Proc.*, pp. 885–890, 2010, doi: 10.1109/IROS.2010.5650283.

[56] G. de Boer et al., "Design optimisation of a magnetic field based soft tactile sensor," *Sensors (Switzerland)*, vol. 17, no. 11, 2017, doi: 10.3390/s17112539.

[57] A. Alfadhel, M. A. Khan, S. Cardoso De Freitas, and J. Kosel, "Magnetic tactile sensor for braille reading," *IEEE Sens. J.*, vol. 16, no. 24, pp. 8700–8705, 2016, doi: 10.1109/JSEN.2016.2558599.

[58] S. Takenawa, "A magnetic type tactile sensor using a two-dimensional array of inductors," *Proc. – IEEE Int. Conf. Robot. Autom.*, no. 1, pp. 3295–3300, 2009, doi: 10.1109/ROBOT.2009.5152420.

[59] A. Alfadhel and J. Kosel, "Magnetic nanocomposite cilia tactile sensor," *Adv. Mater.*, vol. 27, no. 47, pp. 7888–7892, 2015, doi: 10.1002/adma.201504015.

[60] N. F. Lepora and B. Ward-Cherrier, "Superresolution with an optical tactile sensor," *IEEE Int. Conf. Intell. Robot. Syst.*, vol. 2015-Decem, pp. 2686–2691, 2015, doi: 10.1109/IROS.2015.7353744.

[61] J. S. Heo, J. H. Chung, and J. J. Lee, "Tactile sensor arrays using fiber Bragg grating sensors," *Sensors Actuators, A Phys.*, vol. 126, no. 2, pp. 312–327, 2006, doi: 10.1016/j.sna.2005.10.048.

[62] H. Yamazaki, M. Nishiyama, K. Watanabe, and M. Sokolov, "Tactile sensing for object identification based on hetero-core fiber optics," *Sensors Actuators, A Phys.*, vol. 247, pp. 98–104, 2016, doi: 10.1016/j.sna.2016.05.032.

[63] J. C. Yeo, J. Yu, Z. M. Koh, Z. Wang, and C. T. Lim, "Wearable tactile sensor based on flexible microfluidics," *Lab Chip*, vol. 16, no. 17, pp. 3244–3250, 2016, doi: 10.1039/c6lc00579a.

[64] S. Sokhanvar, M. Packirisamy, and J. Dargahi, "MEMS endoscopic tactile sensor: Toward in-situ and in-vivo tissue softness characterization," *IEEE Sens. J.*, vol. 9, no. 12, pp. 1679–1687, 2009, doi: 10.1109/JSEN.2009.2025586.

[65] S. Najarian, J. Dargahi, M. Molavi, and H. Singh, "Design and fabrication of piezoelectric-based tactile sensor for detecting compliance," *IEEE Int. Symp. Ind. Electron.*, vol. 4, no. 424, pp. 3348–3352, 2006, doi: 10.1109/ISIE.2006.296003.

[66] A. M. R. F. El Bab, K. Sugano, T. Tsuchiya, O. Tabata, M. E. H. Eltaib, and M. M. Sallam, "Micromachined tactile sensor for soft-tissue compliance detection," *J. Microelectromechanical Syst.*, vol. 21, no. 3, pp. 635–645, 2012, doi: 10.1109/JMEMS.2012.2184080.

[67] L. Zhang, F. Ju, Y. Cao, Y. Wang, and B. Chen, "A tactile sensor for measuring hardness of soft tissue with applications to minimally invasive surgery," *Sensors Actuators, A Phys.*, vol. 266, pp. 197–204, 2017, doi: 10.1016/j.sna.2017.09.012.

[68] P. Peng, S. Member, R. Rajamani, S. Member, and A. G. Erdman, "Flexible tactile sensor for tissue elasticity measurements," *J. Microelectromechanical Syst.*, vol. 18, no. 6, pp. 1226–1233, 2009.

[69] Y. Hasegawa et al., "A micromachined active tactile sensor for hardness detection," *Sensors Actuators, A Phys.*, vol. 114, no. 2–3, pp. 141–146, 2004, doi: 10.1016/j.sna.2003.11.038.

[70] Y. Hasegawa, M. Shikida, H. Sasaki, K. Itoigawa, and K. Sato, "An active tactile sensor for detecting mechanical characteristics of contacted objects," *J. Micromechanics Microengineering*, vol. 16, no. 8, pp. 1625–1632, 2006, doi: 10.1088/0960-1317/16/8/026.

[71] R. Fernandez, I. Payo, A. S. Vazquez, and J. Becedas, "Micro-vibration-based slip detection in tactile force sensors," *Sensors (Switzerland)*, vol. 14, no. 1, pp. 709–730, 2014, doi: 10.3390/s140100709.

[72] C. M. Oddo, L. Beccai, M. Felder, F. Giovacchini, and M. C. Carrozza, "Artificial roughness encoding with a bio-inspired MEMS- Based tactile sensor array," *Sensors*, vol. 9, no. 5, pp. 3161–3183, 2009, doi: 10.3390/s90503161.

[73] Y. Jiang, Z. Ma, B. Cao, L. Gong, L. Feng, and D. Zhang, "Development of a tactile and slip sensor with a biomimetic structure-enhanced sensing mechanism," *J. Bionic Eng.*, vol. 16, no. 1, pp. 47–55, 2019, doi: 10.1007/s42235-019-0005-8.

4

Optical-based Technologies for Artificial Soft Tactile Sensing

Matteo Lo Preti[1,2], Massimo Totaro[1], Egidio Falotico[2] and Lucia Beccai[1,*]

[1]Center for Micro-BioRobotics, Istituto Italiano di Tecnologia, Pontedera, Pisa, Italy
[2]The BioRobotics Institute, Scuola Superiore Sant'Anna, Pontedera, Pisa, Italy
E-mail: matteo.lopreti@iit.it; lucia.beccai@iit.it
*Corresponding Authors

The field of robotics is in continuous expansion, with application scenarios shifting from structured environments, such as factories, to the more complex ones of the real world, such as houses, hospitals, disaster areas, etc. Artificial tactile sensing is fundamental in this process to enable the full immersion of the robotic systems in the environment making them aware of their surroundings, and capable of reacting to unexpected situations. Optical-based soft artificial skins seem to be a promising solution both to enable advanced sensing capabilities and to overcome some critical integration issues. Indeed, in this case, the sensitive area can be built with skin-like materials free of any active component, allowing a natural mechanical interaction between the system and the environment. This solution introduces the need for more sophisticated signal processing to retrieve useful information. Nevertheless, machine learning algorithms can open the way to real-time reconstruction methods that can also be embedded in systems with low computational capabilities. In this chapter, the recent developments of optical-based soft tactile sensing are described. As a case study, a comparison between machine learning and analytical approaches, to decode tactile information in a continuum soft optical waveguide, is presented.

4.1 Introduction

Artificial tactile sensing is fundamental in many fields, such as robotics, prosthetics, wearable systems, or haptics [1–4]. From a transduction point of view, mainly resistive-and capacitive-based systems were investigated successfully [5]. In the first case, the main advantages are the easy fabrication, the simple read-out circuitry, the very high sensitivity, and the relatively high acquisition rate. In the second case, mainly due to their high linearity, insensitivity to temperature variations, capacitive sensors have become the most used technology for several touch-based human–machine interfaces. This mainly resulted from the impressive advancement of miniaturised electronics for very low capacitance measurement (down to the sub-femtofarad range) [6, 7].

In robotics, several very large area electronic skins have been developed in recent years. Some examples can be found in humanoid robotics, such as those integrating capacitive sensors, in the iCub platform [8], or multimodal patches for temperature, acceleration, and proximity in the H-1 robot [9]. In both cases, a modular architecture is implemented to cover the whole robot surface. Although this implies that a large number of wires and connections is routed through the robots, the final systems can still be reliable, since the robotic structures are built from rigid materials. However, the recent rise of soft robotics [10] introduced new challenges in tactile sensing [11]. Here, previous sensing architectures are not suitable mainly because, both, the rigid (or semi-rigid) modular patches, and the high number of distributed tactile elements (i.e. requiring a high number of wires/connections), would introduce undesired mechanical constraints in the soft robot and embedded sensing systems. Therefore, new sensing strategies for soft perceptive robots should be developed.

Among other principles, optical-based tactile sensing has shown a relevant boost in recent years. From a transduction point of view, using light minimises the influence of electromagnetic noise, thus allowing its use in a large number of scenarios, including the harshest ones, e.g., surgery or harsh environments. Particularly in the case of the challenging large-area sensing, the main strengths of the optical-based approach are high spatial resolution, reliability, and improved frequency rate. Noteworthy, in extended optical waveguides, the area in which the tactile transduction occurs does not require a distributed array of active electronic components; instead the latter (e.g. emitters and receivers) can be located at the periphery of the sensitive area. Moreover, the integration of such active elements is easily enabled due to highly efficient and miniaturised electronic components – such as light-emitting diodes (LEDs), charge-coupled devices (CCDs), photodiodes, and phototransistors. On the other hand, light along a waveguide undergoes small losses due to microbending and the overall signal can be altered by the misalignment of active optical devices. Also, the reconstruction process is complex, since there is no active element within the sensing area, and smart processing techniques need to be used.

This chapter will start with some basic physical principles underlying optical-based sensors (Section 4.2) and by proposing some key aspects based on which they can be classified (Sections 4.3). Then, the most important ways to process data coming from these sensors will be reported, and a case study concerning a soft optical waveguide skin will be discussed (Section 4.4). Finally, in the conclusion part, the main aspects concerning the suitability and bottlenecks of existing processing techniques for such kind of optical sensors will be discussed (Section 4.5).

4.2 Optical-based Tactile Sensors

An optical-based tactile sensor is a device or system that can decode a mechanical stimulation (i.e. pressure, strain, bending, twisting, etc.) by exploiting light properties, e.g., intensity, wavelength, coherence, phase, and polarisation. Most generally, electromagnetic waves in the optical spectrum are guided by a physical structure, called *optical waveguide*, which acts both as a light transmission medium and mechanical transducer. Among them, the most common types include optical fibres and transparent dielectric waveguides.

The following characteristics can be effectively used to distinguish between different kinds of optical sensors [12]:

- *Sensing extent*: there exists point sensors, which have a one-to-one correspondence between emitters and use phase change for the activation, and distributed sensors, whose sources can stimulate the receptive area of more than one receptor;
- *Transduction location*: two main cases can be identified: (1) Extrinsic sensor, in which the light beam leaves the optical waveguide, and its properties vary before reaching an optical detector, e.g. due to an external object in its path; (2) Intrinsic sensor, in which the light beam variation, detected by the optical receiver, occurs inside the waveguide, e.g. if it deviates in response to a mechanical stimulation of the waveguide itself.

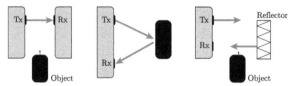

Figure 4.1   Different kinds of optical sensors: Through beam (left), reflective (centre), and retro-reflective (right).

- *Sensing method* and *placement* of optical transmitter and receiver (Figure 4.1):
  - *Through beam*: transmitter and receiver point to each other so that they create a straight light beam path. When any object comes in between, the intensity of light changes and the object can accordingly be detected;
  - *Reflective*: transmitter and receiver are parallel to each other. The object reflects light[1] from the transmitter, and the receiver measures it;
  - *Retroreflective*: the reflector both reflects and receives light coming from the transmitters and the receivers [13].

4.2.1 Basic Optical Principles

In an optical waveguide, a medium in which light can travel with a negligible loss is employed [14]. This represents the core material that in a fibre is totally surrounded by a cladding, whereas in a planar waveguide is interfaced to an external medium at the top and bottom surfaces. As it will be explained later, to guide the inserted light efficiently (negligible light loss) in the optical device, the optical properties of the core material must be such that its refractive index is of higher value than for the cladding or external medium.

The index of refraction $n = v/c$ describes the behaviour of the light across a medium. Indeed, for an external observer, an electromagnetic wave travels within a dielectric medium at a speed v lower than c, which is the speed of light in the vacuum [15].

The principle is regulated by Snell's law, according to which *the ratio between the sine of the incidence and refraction angles is equal to the relative*

---

[1]This type of sensor has a drawback to differentiate between red and white light when red colour LED is used as an optical source. This is because both red and white colours have the same amount of reflection.

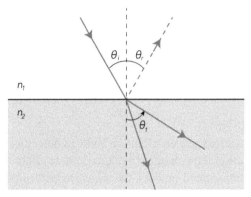

Figure 4.2 Reflection and refraction of an electromagnetic wave.

refractive index between the two media. In particular

$$n_1\sin\theta_1 = n_2\sin\theta_2$$

Considering the case of reflection on specular surfaces, i.e., a near-perfect flat surface, after reflection from a specular surface, light rays are parallel to each other and are heading in the same direction. Therefore, the wavefronts remain unchanged.

As shown in Figure 4.2, when a light wave (represented as a red line) runs from a medium whose index of refraction is n_1 towards a medium with index n_2, it can either increase or decrease the angle with respect to the normal to the surface:

$$\sin\theta_2 = \frac{n_1}{n_2}\sin\theta_1 \Rightarrow \begin{cases} \theta_2 < \theta_1 & n_2 > n_1 \\ \theta_2 > \theta_1 & n_2 < n_1 \end{cases}$$

The second case ($\theta_2 > \theta_1$) has a case limit when $\theta_2 = \pi/2$, which corresponds to the incidence angle, $\sin\theta_0 = n_2/n_1$.

If $\theta_1 > \theta_0$, no refracted wave is produced, and the original wave is totally reflected in the first medium. Then, in the case of the fibre, if the refractive index of its core (n_1) is higher than that in the cladding (n_2), then total internal reflection can occur, and all the rays inserted in the fibre with angles higher than θ_0 will reach the other end of the fibre.

4.2.2 Pressure and Strain Optical Sensing

Optical waveguide sensors can also be classified by their working principles. In particular, in the following, pressure and strain sensors will be considered, as pressure and strain are two important touch-related parameters.

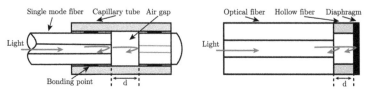

Figure 4.3   Typical structure of EFPI-based pressure sensor (left) and DEFPI-based pressure sensor (right).

Optical pressure sensors use an effect on light to detect a change in pressure. In the simplest case, a mechanical system deviates light as the pressure increases. In more sophisticated sensors, the measurement of phase difference sharply increases the accuracy, since a reference photodiode – which is always reached by the light – is used to correct for changes in the output due to factors like the ageing of the light source, or variations in supply voltage [16]. Other optical sensors use interferometry to measure the phase change between the light that has taken two different optical paths, corresponding to a fraction of the wavelength of light. In particular, interferometry is used in fibre pressure sensors, namely the extrinsic Fabry-Perot interferometer (EFPI) and fibre Bragg grating (FBG).

The extrinsic type sensor uses interference between multiple light rays reflected back and forth between two mirrors at opposite ends of a cavity. They transmit a fraction of the resonant frequency by making the distance of one round trip between mirrors equal to an integer number of wavelengths of the cavity material. Hence, the amount of light received at a particular wavelength changes as the spacing changes.

A typical EPFI system is schematically represented in Figure 4.3, and it includes a light source, a spectrometer, an optical coupler, and a sensor head, which can be a capillary-based EFPI structure for static and high-pressure measurements, or a diaphragm-based EFPI (DEFPI) structure for low-pressure range and dynamic measurement [17]. The source emits the light into a fibre coupler and propagates along the lead-in/out fibre to the sensor head. A small part of the incident light is firstly reflected and then returns directly back to the fibre. The remaining part of the light propagates across the air gap to the end-face of the reflecting fibre, and then the reflected light is recoupled into lead-in/out fibre and interferes with the first reflected light. The output interference signal can be expressed by [18]

$$I = 2I_s\left(\lambda\right)\left(1 + \gamma\cos\varphi\right)$$

where $I_s(\lambda)$ is the intensity distribution of the light source, γ is the fringe visibility of EFPI, $\varphi = 4\pi d\ \lambda$ is the round trip, and d is the cavity length.

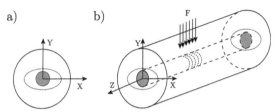

Figure 4.4 (a) Cross section of the FBG with marked fibre shape (dashed line) under pressure. (b) Schematic drawing of the FBG with the indicated direction of applied force F [19].

For the sensor geometry shown in Figure 4.3 (left), a pressure P on the sensor head results in a longitudinal compression of the alignment tube, and, consequently, in a change in the cavity length of EFPI, resulting in the following equation:

$$\Delta d = \frac{Lr_0^2}{Er_0^2 - r_i^2}\left(1 - 2\mu\right) P$$

where E is the Young's modulus of the tube, μ is the Poisson ratio, L is the distance between two thermal fusion points, r_o and r_i are outer and inner radii of the capillary tube, respectively. These equations reveal that the pressure acting on the EFPI sensor changes linearly with respect to the cavity length. Such pressure can be retrieved by analysing variations of the interference spectrum by means of some specific signal processing algorithms.

An FBG is an intrinsic sensor that has a regular series of reflective structures in the fibre; the wavelengths of the reflected light change by stretching or compressing the fibre itself (Figure 4.4).

The FBGs are formed by a periodic change of the fibre core refractive index in the direction of propagation of optical radiation. In principle, the FBG acts as a spectral filter that reflects particular wavelengths of light near Bragg resonance wavelength, and the rest of the optical signal spectrum is released. The Bragg resonant wavelength is given by

$$\lambda_{\text{Bragg}} = 2n_{\text{eff}}\Lambda$$

where λ_{Bragg} is the Bragg resonant wavelength, n_{eff} is the effective refraction index, and Λ is the periodic variation of the FBG. The FBGs used in sensors mostly rely on the spectral analysis of the reflected light wavelengths [19].

Alternatively, tactile sensors can rely on total internal reflection (TIR), which is indeed total if the external medium is perfectly transparent, continuous, and of infinite extent. All the other cases result in some losses, e.g., the

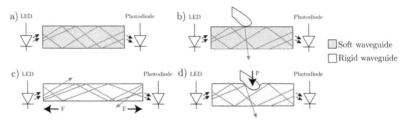

Figure 4.5 Light transmission in a rigid or soft waveguide: (a) without interferences, (b) FTIR in a rigid waveguide caused by finger touch, (c) a soft waveguide is stretched, and (d) FTIR in a soft waveguide to which a pressure is applied.

evanescent wave carried by the external medium can be absorbed by a lossy external medium (*attenuated total reflectance*) or diverted, either by its outer boundary or by objects embedded in that medium (*frustrated* TIR or FTIR). Considering the case of a planar waveguide, as schematised in Figure 4.5(a), a light beam is injected into the waveguide, and, being confined by TIR, it can reach the detector without losses. On the other hand, as schematised in Figure 4.5(b), FTIR causes a change of the detector output current when, for example by finger touch, one medium (i.e. air) is substituted by another (i.e. finger). Indeed, at a given contact location, the angular condition for TIR is no longer satisfied, and light is partially transmitted from the opposite side of the waveguide.

In both the abovementioned cases, the waveguide is a rigid material that does not deform upon mechanical stimulation. However, a soft, optically clear material can be used as a waveguide medium. In this case, an exerted pressure (Figure 4.5d) or traction (Figure 4.5c) causes deformation of the waveguide surface, or an overall stretching condition, respectively. In both the cases, light losses occur, since the internal reflection conditions are varied locally (pressure) or globally (stretching), and the critical incidence angle of some rays is overcome. In recent years, exploiting this concept, new designs for soft tactile sensing were introduced, as it will be described in the following sections.

4.3 Examples of Optical-based Tactile Sensors

4.3.1 Single Optical Waveguide Sensor

Several optical-based soft tactile sensors rely on the deformation of an optical waveguide, built with an optically clear soft material shaped in a long and narrow lace-like structure[2] that have an emitter (i.e. photodiode) and a receiver (i.e. phototransistor) at each extremity. The variation of light intensity detected by the receiver is correlated to a mechanical deformation, due for instance to an externally applied pressure, or to the strain induced by elongation or bending. Figures 4.5(c) and 4.5(d) show a simple schematisation of the principle when strain is induced or pressure is applied to a deformable waveguide, respectively. Moreover, stretchable fibres are developed by mechanically coupling a highly deformable transparent cylindrical core and a surrounding soft cladding.

A typical material in which emitters and receivers are integrated is polydimethylsiloxane (PDMS), which is a soft, conformable, and compliant material, optically transparent for wavelengths ranging from 235 nm to the near-infrared and whose attenuation is as low as 0.4 dB/cm.

For instance, in Ref. [20], stretchable optical fibres are sewed in textiles for strain measurement. In this way, they can detect strains originating from changing muscle shapes during weight-bearing activity. Their core is built from urethane fibre, and they have a silicone cladding. They can stretch up to 100% of their original length, with negligible hysteresis. Similarly, a strain sensor for human motion detection is reported in Ref. [21]. The stretchable fibre is fabricated from dye-doped PDMS, and the sensor can detect strains over 100% with about 1% precision. In this case, PDMS is doped with rodhamine particles to enhance the light attenuation due to stretching. As application examples, the sensor is demonstrated to detect dynamic motions of the body, such as joint motion, as well as vibrations due to speaking or deep breathing in real time.

An interesting application of stretchable lace waveguide is reported in Ref. [22], where curvature, elongation, and force were detected in a nylon fibre-reinforced artificial soft finger. In this case, three lace waveguides were integrated into each finger, where each finger is bent into a U-shape in a way that light is not heavily attenuated (Figure 4.6b). When the soft finger is bent, the waveguide located at the top (convex part) detects the largest axial strain with respect to the other two, i.e. the second one integrated at the finger middle plane and the third one at the finger inner part (concave part), and thus it has the largest sensitivity to the bending motion. The

---

[2]In the artificial tactile sensing literature, a terminology that is often used for this kind of devices is *optical fibre*. Nevertheless, in optoelectronics an optical fibre is a cylindrical structure with well-defined core and cladding. Therefore, for waveguides that are long and narrow, but not constructed like typical optical fibres, the name that will be used here is optical lace waveguides.

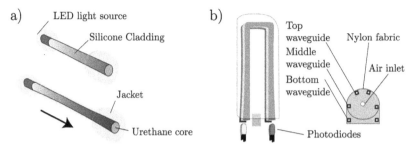

Figure 4.6   (a) Change in light attenuation along a stretched urethane fibre [20] and (b) front view and cross section of three soft lace waveguides integrated into an artificial finger shape [22].

second waveguide conveys information about internal pressure, while the third waveguide serves as a touch sensor for the fingertip of the prosthetic hand. This isolated function is obtained by placing it at the neutral bending plane, where there is no axial strain.

4.3.2 Bundle Optical Waveguide System

The development of systems embedding a bundle of fibre- or lace-like waveguides can allow the detection of different mechanical stimulations at the same time. For instance, in Ref. [23] an internally illuminated elastomeric foam is reported, as schematised in Figure 4.7(a). The waveguides have the role of, both, transmitting light into the foam, and receiving diffuse waves (from internal reflection) that are interpreted by machine learning techniques to predict whether the foam is twisted clockwise, counterclockwise, and bent up or down (also giving the magnitude of the deformation type). The model predicts the type of deformation with 100% accuracy, and of magnitude with a mean absolute error of 0.06°.

Another example is given by stretchable lace waveguides distributed throughout a 3D elastomeric framework [24] (Figure 4.7b), which created a co-integrated body, sensing, and communication network. In particular, a scaffold made of 3D printed elastomeric material is used to house a four-cored optical lace. In each lace, the light enters in an input optical waveguide, which is mechanically coupled with different output waveguides. When a mechanical stimulation occurs, the light is coupled differently in each output waveguides. Then, by machine learning techniques, both deformation and force with sub-millimetre and with sub-Newton resolution can be detected, respectively.

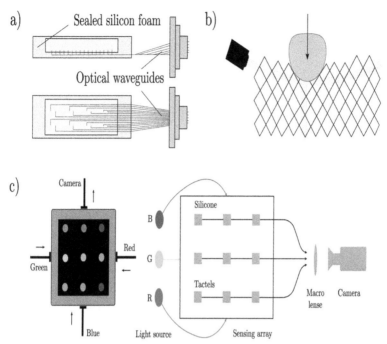

Figure 4.7 (a) Foam and optical waveguide assembly in Ref. [23], (b) schematic showing how light (yellow) is coupled to an output core when the 3D structure with soft lace waveguides is deformed [24], and (c) colour-coded tactile sensor: assembly with LEDs, camera, and plastic optical fibres, and camera snapshot overlaid onto the sensor. The camera image is processed to infer the sensor deformation. The block diagram illustrating the sensing principle of the colour-coded optical tactile sensor [25].

On the other hand, optical fibres were utilised in the approach of Kappassov et al. [25], for developing a colour-coded tactile sensor. Three sources deliver light via plastic optical fibres (POFs) embedded inside a transparent silicone layer to an off-the-shelf camera (Figure 4.7c). When an external object comes in contact with the silicon and compresses it, the light scattering pattern changes. In particular, a change in the colour is used to determine the depth level at which the silicone is deformed. Thus, the colour-coded silicon substrate acts as a pressure-sensing media and changes colour to signal pressure level. In this work, all the processing electronics is located away from the sensing surface in order to promote integration in various scenarios such as robot grippers to improve their dexterity in physical interaction tasks, or tactile motion control (e.g. squeezing an object for determining its deformability).

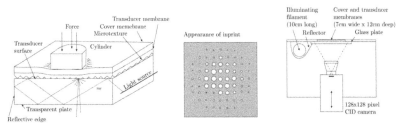

Figure 4.8   Large-area TIR force and pressure sensor and diagram illustrating how areas subject to higher pressures appear as regions of higher light intensity [27].

4.3.3 Continuum Optical Waveguide Skin

A continuum optical waveguide skin can be obtained if an extended layer of optically clear material is used as an optical waveguide; thus, the concept exemplified in Figure 4.5(a,b) is applied to an extended area. Following this approach, the distribution of tactile information (e.g. contact, pressure, etc.) over a large area can be obtained from a sensitive region which, in principle, can be totally free from any electronic components and wiring.

One of the first works that exploited the phenomenon of TIR in a large-area tactile sensor is that of Begej et al. [26]. For instance, Figure 4.8 shows a diagram of a large-area tactile sensor utilizing a lens-based imaging system. A 7 × 12 cm window glass is used as a transparent layer. The glass is coated with a Sorta Clear silicone layer with Shore level 18 (transducer membrane). Over this, another Sorta Clear 18 layer mixed with nickel is added (cover membrane). The metallic powder increases the reflectance of the silicone compound. Light sources, guided by POFs, are injected in the sensing silicone layer. Then, a camera captures the imprint image in the silicone, and the recorded signals are processed by machine learning classification algorithms. The result is a map containing the location and amplitude of normal applied force.

A continuum optical waveguide skin can be implemented with soft optically clear materials (see concept exemplified in Figure 4.5(c,d)). Indeed, a distribution of emitters and detectors can be integrated at the periphery of the sensitive area in layouts that allow for the reconstruction of the tactile map. The possibility to rely on the entire substrate as a continuum waveguide for large-area sensing can contribute to overcome some of the typical issues related to using bundles of fibres, i.e. crosstalk, complex shielding, and fabrication. Indeed, meaningful spatial information can be extracted from optical data by developing new processing strategies, e.g. machine learning techniques.

This approach is very attractive not only from a robotic point of view, but also for any human–machine device, where tactile sensitive areas must be integrated into complex 3D and actuated parts. These challenges are even more complex in soft robotics, where highly deformable and morphing structures are developed [11, 26].

However, the advantages of a decreased sensing layout complexity are counterbalanced by the need of complex reconstruction processes, as better detailed in the next section, and as represented by some examples [27, 28]. In the first case, a 150 μm thick planar waveguide is used as flexible force sensor array. The waveguide is made of two photocurable fluorinated prepolymers, and inside the waveguide, cylindrical spacers (having a diameter of 200 μm) are microstructured in order to define each taxel. The force sensor array detects a contact force in 27 points independently in a 0–3 N range without any electronic components on the sensing area. In the second case, a soft optical waveguide skin integrating eight infrared emitters and eight photodetectors at the periphery of a PDMS layer is reported. Pressure up to 160 kPa is measured with a spatial resolution in the centimetre scale in the whole sensitive area.

4.4 Signal Processing Approaches for Continuum Optical Waveguide Skins

Typically, to retrieve the necessary spatially resolved tactile information in an artificial skin consisting of an array of sensors, the signals coming from the sensors need to be processed in ways to extract meaningful information correlated to the tactile cues.

Regarding continuum optical waveguide skins, several approaches for the decoding algorithms are possible, such as backprojection [29], machine learning [30], or the neuromorphic approach, in particular for what concerns the field of the neural interfaces [31].

In this kind of sensing, the most difficult challenges concern the number of emitters that can elicit the same receiver, and how fast the received signal stabilises to informative values. Since there is not a one-to-one correspondence between emitters and receivers, even small changes in the input may correspond to very different outputs.

4.4.1 Analytical Methods

Usually in large-area electronic skins, typically two main approaches have been used. In the first case, the major complexity burdens on the hardware architecture, relying on an *array with row/columns addressing*. Oppositely, using an isotropic conductive material with an array of electrodes at its boundary, *electrical impedance tomography* (EIT) algorithms have been developed. The former is simple, fast, it has low-cost read-out electronics, but its spatial resolution is limited, and it needs complex wiring/connections. On the other hand, the latter has higher spatial resolution and simpler wirings, even though the computational cost is higher, and the maximum achievable frequency is lower.

EIT is an approach originally developed for resistive-based electronic skins, but equivalent algorithms can be widely adopted for other kinds of skins. Some highlights may allow to clarify some of the problems related to the reconstruction processing. This technique reconstructs the conductivity distribution of a domain under test [32]. The theory behind the EIT is that a constant current across a material results in a voltage distribution on the surface that will reflect the internal resistivity distribution. However, multiple resistivity distributions can produce the same superficial voltage distribution. Therefore, this problem is solved by stimulating the system in multiple manners to constrain the possible resistivity distributions [33]. The reconstruction problem consists of obtaining an approximation of the internal area from the boundary measurements, which is both highly nonlinear and ill-posed, meaning that large changes in the interior can correspond to minimal changes in the measured data. The different approaches fall into several categories, such as linear approximations, iterative methods, adaptive methods, and the so-called layer-stripping algorithm [34].

Regarding soft continuum optical waveguide skins, a tomography-like approach was applied for the first time in Ref. [28] (Figure 4.9).

The light signal is converted into an electrical signal through an electronic system, and a reconstruction algorithm similar to tomographic back projection reconstructs the pressure map using the following relation:

$$P(i,j) = F\left(\sum_{i=1}^{N} \sum_{k=1}^{M} J_i\, J_k\right)$$

where P is the pressure applied on the waveguide in the points (i,j) of the surface, J is the light intensity collected by the detectors, and the function F implements trapezoidal interpolation to the pieces of information in the sensing area.

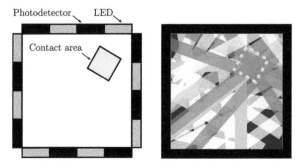

Figure 4.9 Schematisation of the tomography-like approach in the sensor developed by Levi et al. [28].

4.4.2 Machine Learning Methods

Machine learning is an important possibility to process data in optical-based sensors (such as in other types of artificial tactile skins), and it consists of the automated process of pattern extraction from data. In predictive data analysis applications, supervised learning is commonly used to automatically learn a mapping between a set of instances given by the combinations of descriptive feature values and the target feature to obtain a model. However, this approach represents an ill-posed problem, since the best model cannot be found based on the sample training dataset alone. Indeed, it must be endowed with generalisation capabilities, i.e., it must able to predict queries not present in the dataset; thus, some criteria are needed to choose among the candidate models [35].

Classification is the process of finding a model (or function) to describe and distinguish categorical data classes, based on the analysis of a set of data objects for which the class labels are known. Since the class label of each training tuple is provided, this step is also known as supervised learning. Data classification is a two-step process, consisting of a learning step (or training phase, where a classification model $y = f(X)$ is constructed, that can predict the associated class label y of a given tuple X) and a classification step (or test phase, where the model is used to predict class labels for a given data, and its performance are evaluated in terms of a given metric) [36, 37].

It is useful to give an overview of the main categories of classification techniques, as it follows:

- *Decision Tree Induction*: a decision tree (Figure 4.10a) is a flowchart-like tree structure, where each internal node (non-leaf node) denotes a test on an attribute, each branch represents an outcome of the test, and

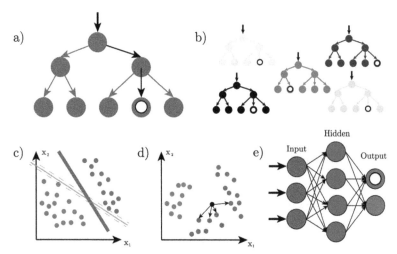

Figure 4.10   Examples of (a) a decision tree, (b) random forest, (c) SVM, and (d) kNN, neural netork for classification with one hidden layer.

each leaf node (or terminal node) holds a class label. Most algorithms (e.g. ID3, C4.5, and CART) adopt a greedy top-down recursive divide-and-conquer manner to build the tree. Then, given a test tuple, its attribute values are tested against the decision tree. A path is traced from the root to a leaf node, which holds the class prediction for that tuple [38];

- *Ensemble Methods:* Random Forest (Figure 4.10b) is an example of an ensemble method, which combines a series of learned models (or base classifiers) to create an improved composite classification model. When a new data to classify is provided, the class prediction results from the combination of the output of the single classifiers according to a chosen metric, e.g. majority voting, Random forest is quite robust to errors and outliers. The generalisation error for a forest converges as long as the number of trees in the forest is large. Thus, overfitting is not a problem. The accuracy of a random forest depends on the strength of the individual classifiers and a measure of the dependence between them [39];

- *Support Vector Machine:* Support Vector Machines (SVMs, shown in Figure 4.10c) use a non-linear mapping to transform the original training data into a higher dimension [40]. Within this new dimension, it searches for the linear optimal separating hyperplane (i.e. a decision boundary separating the tuples of one class from another). With an appropriate non-linear mapping to a sufficiently high dimension, data from two classes can always be separated by a hyperplane. Furthermore, the so-called Kernel trick consists in mapping the input space in a higher dimensional space by using kernels. Indeed, data that are not linearly separable in input space, may yet become linearly separable in the higher dimensional feature space [41]. Although the training time of even the fastest SVMs can be extremely slow, they are highly accurate, owing to their ability to model complex non-linear decision boundaries. They are much less prone to overfitting than other methods;

- *Lazy Learners*: k-Nearest-Neighbour (kNN, Figure 4.10d) is a lazy learner based on learning by analogy. This means that it stores training tuples and waits until it is given a test tuple. Only when it receives a test tuple, it performs generalisation to classify the tuple based on its similarity to the stored training tuples. For the kNN classification, the unknown tuple is assigned the most common class among its kNNs. Hence, when making a classification or numeric prediction, lazy learners can be computationally expensive. However, they naturally support incremental learning [42];
- *Neural Networks*: a neural network (Figure 4.2(e)) is a set of connected input/output units in which each connection is associated with a weight. During the learning phase, the network learns by adjusting the weights to predict the correct class label of the input tuples [43]. Before training can begin, the user must decide the network topology by specifying the number of units in the input layer, the number of hidden layers, the number of units in each hidden layer, and the number of units in the output layer. Neural networks involve long training times, they require some parameters that are typically best determined empirically, such as the network topology and they have often been criticised for their poor interpretability. Advantages of neural networks, however, include their high tolerance of noisy data as well as their ability to classify patterns on which they have not been trained. They can be used when there is little knowledge of the relationships between attributes and classes. They are well suited for continuous-valued inputs and outputs.

For instance, in Ref. [30], a processing methodology for tactile data obtained by means of a soft continuum optical waveguide skin (Figure 4.11) is presented. Indeed, the drop in light intensity, measured by photodetectors

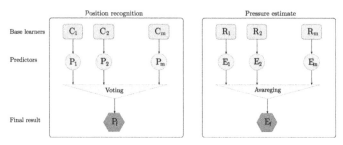

Figure 4.11   Machine learning approach for single touch offline detection in a soft optical waveguide skin [30].

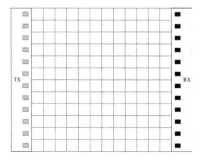

Figure 4.12   Scheme of the optical sensor developed by Borg et al. [44] (with emitters [TX] on the left and receivers [RX] on the right) which reaches 0.57 cm of accuracy for the position retrieval and a force level to within 1.37 N of accuracy.

when indentations are performed, is correlated to the tactile sensor's response using adaptive boosting and decision trees.

Another example that can be considered is the recent work of Borg et al. [44], in which a flexible and transparent gel slab is employed as an optical artificial skin, providing white light from one side and measuring the amount of light received on the opposite side with phototransistors to retrieve position and pressure information, as it is shown in Figure 4.12.

Data from the phototransistors are processed by means of an adaptive neuro-fuzzy inference system technique (ANFIS) that refers to a class of adaptive networks that are functionally equivalent to fuzzy inference systems (FIS) [45]. Fuzzy logic is suitable for dealing with ill-defined and uncertain systems. The FIS systems are used to map inputs to outputs relying on fuzzy rules, which are linguistic *if-then* statements that describe how decisions should be made. The combination of both fuzzy logic and neural networks gives advantages of constructing an input–output mapping based on both user input and stipulated input–output data pairs [46].

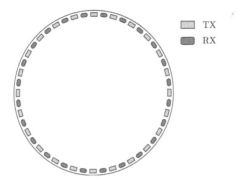

Figure 4.13 Schematic design of the sensor: the circular structure is the PDMS layer, along whose border 24 photodiodes (TX) and phototransistors (RX) have been alternated.

4.4.3 Case Study: Distributed Mechanical Sensing in a Soft Optical Skin

In this section a case study is introduced as an example of development of soft continuum optical waveguide skin addressed by the authors, in which distributed contact and pressure information is retrieved by means of low computational strategies [47]. The skin is built from a round soft polymeric layer in which light (in the infrared spectrum) is injected and decoded by means of an array of photodiodes and phototransistors located in an alternate fashion at the edge of the sensing area, as shown in Figure 4.13.

This choice is fundamental for two main reasons, i.e. the sensing area is left free of any sensing element, whose number scales with the radius and not with the area of the surface.

In addition, high number of electronic components cause usual issues of high power consumption. To reduce the amount of energy needed to power up the skin one photoemitter at a time has been activated, in a circular fashion. This causes a time delay of external read-out circuitry, since the photoemitters must be activated for the time needed to guarantee a stable optical signal. However, a good trade-off can be found by minimizing such activation time.

According to the optical principles described in Section 4.2, the deformation of the soft waveguide induced by an externally applied mechanical stimulus, can be measured as a variation of the light intensity that arrives at the phototransistors and, accordingly, as a variation of their output current.

As a first step, a mechanical characterisation is needed to quantify the dependency between the optical signal (explanatory variable) and the force applied upon the surface of the sensor. Once this relationship is established,

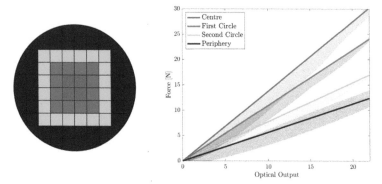

Figure 4.14 Linear fitting for the force vs optical output curves for the response of the sensor going from the centre towards its periphery. The error comes from the average between all the cells belonging to the same category, i.e. centre, first circle, second circle, and periphery of the sensing area.

an estimate of the contact area serves straightforward the conversion into pressure. This is carried out by means of an electromechanical characterisation setup, in which an external force is applied by means of an indenter having its position controlled in the direction perpendicular to the artificial skin surface. To obtain the reference force values, a three-axis load cell is utilised to record the indentation forces at predetermined indentation displacements. Figure 4.14 shows the linear fitting curves and their respective error with respect to the true force values.

It can be noticed that the optical response does depend on the distance with respect to the centre of the sensing area. Indeed, the different zones (highlighted in Figure 4.14 with different colours) need different scaling factors to infer the overall force exerted upon the surface given the optical signal. Hence, force estimate and spatial position recognition must be processed separately. In particular, a comparative machine-learning-based processing can be adopted and tested to retrieve the information about single-touch contacts. Among the strategies reported in Sections 4.1 and 4.2, a back-projection-like approach and machine learning techniques will be now discussed.

The former method is based on the fact that each photodiode emits light within a cone of emission that allows to identify which phototransistors will be hit. Accordingly, some pairs of diode-receivers couples can be highlighted, such that the path that connects them corresponds to a portion of the sensing area. Hence, a change in the response of that couple can be associated

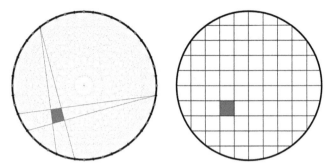

Figure 4.15 Division based on the interaction between photodiodes and phototransistor used in the back-projection-like approach (left) and grid division into cells used for classification (right).

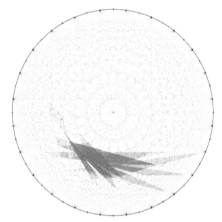

Figure 4.16 Offline reconstruction of data obtained from a single touch using a back-projection-like approach.

with a change in that portion of the surface of the sensing area. Moreover, considering two of these couples at the time, the combined values of their phototransistors are related to the intersection between the two paths (red area in Figure 4.15, left). By considering all the possible intersections, the overall grid can be obtained. However, the extremely large number of intersections represents a limit for the performance of this approach. Figure 4.16 shows the result of an offline reconstruction of a single touch.

Instead, undertaking the machine learning approach, the sensing area can be split, e.g., into a grid like the one reported in Figure 4.15 (right). Consequently, the recognition of the elicited cell can be discretised and solved as a

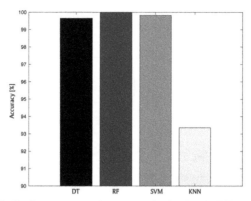

Figure 4.17   Performances on the test set obtained with different classifiers.

classification problem. In this case, a dataset can be built by stimulating each cell with different pressure levels. The output signal from the phototransistors represent the input variables, whereas the elicited cells are used as target. All the elements in the dataset that correspond to the smallest indentation possible can be assigned to class 0, meaning that nothing is pressing on the surface of the sensitive area. In this case study, the equipment used was such that the minimum displacement of the indenter that could be controlled corresponded at 0.1 mm.

Four supervised learning strategies (i.e. decision tree, random forest, SVM, and kNN) have been used, and the result of the 10-fold cross validation training is shown in Figure 4.17: random forest outperforms all the competitors by reaching a perfect accuracy.

This result confirms the idea that the light path is interrupted at different locations when touching different areas of the sensing surface. This phenomenon is particularly clear when the applied force is high. Indeed, it is for low force levels that most of the tested classifiers misclassified the target class. Yet, the overall performance is reasonably good.

4.5 Conclusion

New optical sensing strategies are likely to stimulate several disciplines, and to represent an attractive field of research for many applications, e.g. robotics and biomedical systems [48]. This is especially due to the immunity to electromagnetic interference (EMI) and real-time transduction. In this work, a focus was set on analysing soft optical waveguides, which can mimic the natural skin softness. On the other hand, an accurate choice of the waveguide materials is needed for limiting light losses, which can also occur due to possible deformations of the soft materials themselves, (such as bending, torsion, etc.).

New continuum optical waveguide skins can be designed, providing new exciting solutions for large-area tactile skins, with high spatial resolution, flexibility, and absence of cumbersome interconnections. For this to happen, smart and new processing techniques are needed.

Among the described approaches, back projection has fundamental limitations of high computational complexity, whereas, smart and simple methods are needed to fulfil the online data processing requirement of the majority of applications (robotics in particular). Moreover, within this scenario, as the number of optical components increases, the number of the signals to be processed increases too, making back projection-based algorithms unfeasible, for highly resolved tactile information. In addition, the adaptability of the chosen processing technique may vary with the geometry of the sensor and, accordingly, to different configurations of emitters and receivers.

On the other hand, by adopting a machine learning approach, different classifiers should be trained for different geometries. Nevertheless, it is not unlikely that a parametrisation exists, such that a preliminary recognition of the arrangement of the optical components would allow tuning the model automatically, as a preliminary step to project data onto a shape-independent space. In other words, the algorithm can be trained to recognise a given number of recurrent shapes, and adjust its parameters accordingly, without the need to change the subsequent steps of the processing phase. However, classification techniques have the intrinsic need to have correctly labelled data to be trained with. For example, the grid reported in Figure 4.15 (right) has to be chosen and designed a priori to teach the algorithm the correct mapping. The systems trained with machine learning techniques are in fact opaque, since they behave like black boxes when inputs are provided, whereas analytical algorithmic procedures can be followed step by step. Moreover, machine learning algorithms can only model what they are told to, and their output will inevitably reflect that. In other words, such approach lacks intuition, which can be acquired from a broad learning experience. Nevertheless, machine learning holds high potential for the development of a new generation of perceptive soft robotics [49, 50] and in particular for developing distributed optical sensing technologies areas, since it allows to discover hidden features within potentially poor quality raw sensory data, i.e. quantity rich, yet information poor data.

Following this approach, it can be envisaged that the development of large-area tactile skins in future could rely less on designing a distribution of highly specialized taxels (in terms of materials and fabrication technologies). Rather, the utilization of known optical principles with new soft materials and simple layouts, can generate complex sensory signals that, if combined with AI strategies, could provide complex tactile information in an effective way over larger areas.

Acknowledgment

This project has received funding from the European Union's Horizon 2020 programme under grant agreement No 863212 PROBOSCIS.

References

[1] H. Yousef, M. Boukallel, and K. Althoefer, "Tactile sensing for dexterous in-hand manipulation in robotics – A review," *Sensors Actuators A Phys.*, vol. 167, no. 2, pp. 171–187, 2011.

[2] L. Zou, C. Ge, Z. J. Wang, E. Cretu, and X. Li, "Novel tactile sensor technology and smart tactile sensing systems: A review," *Sensors (Switzerland)*, 17.11: 2653, 2017.

[3] H. Culbertson, S. B. Schorr, and A. M. Okamura, "Haptics: The present and future of artificial touch sensation," *Annu. Rev. Control. Robot. Auton. Syst.*, 2018, 1: 385–409.

[4] A. Nag, S. C. Mukhopadhyay, and J. Kosel, "Wearable flexible sensors: A review," *IEEE Sensors Journal.* 2017, 17.13: 3949–3960.

[5] Y. Wan, Y. Wang, and C. F. Guo, "Recent progresses on flexible tactile sensors," *Mat. Today Phys.* 2017, 1: 61–73.

[6] B. George and V. J. Kumar, "Analysis of the switched-capacitor dual-slope capacitance-to-digital converter," *IEEE Trans. Instrum. Meas.*, 2010, 59.5: 997–1006.

[7] K. B. Balavalad and B. G. Sheeparamatti, "Sensors & transducers: A critical review of MEMS capacitive pressure sensors," *Sens. Tranducers*, 2015, 187.4: 120.

[8] A. Schmitz, P. Maiolino, M. Maggiali, L. Natale, G. Cannata, and G. Metta, "Methods and technologies for the implementation of large-scale robot tactile sensors," *IEEE Trans. Robot.*, 2011, 27.3: 389–400.

[9] P. Mittendorfer and G. Cheng, "Humanoid multimodal tactile-sensing modules," *IEEE Trans. Robot.*, 2011, 27.3: 401–410.

[10] D. Rus and M. T. Tolley, "Design, fabrication and control of soft robots," *Nature*. 2015, 521.7553: 467–475.

[11] H. Wang, M. Totaro, and L. Beccai, "Toward perceptive soft robots: Progress and challenges," *Adv. Sci.*, vol. 5, no. 9, p. 1800541, 2018.

[12] D. Ahuja and D. Parande, "Optical sensors and their applications," *J. Sci. Res. Rev.*, vol. 1(5), pp. 060–068, 2012.

[13] "Overview of Photoelectric Sensors | OMRON Industrial Automation." [Online]. Available at http://www.ia.omron.com/support/guide/43/intr oduction.html [accessed 17-Jan-2020].

[14] D. J. Griffiths, "Introduction to electrodynamics, Griffith-3ed.pdf," *Int. J. Neural Syst.*, 2010.

[15] P. Mazzoldi, M. Nigro, and C. Voci, *Elementi di Fisica. Elettromagnetismo e Onde*, II. EdiSAS, 2008.

[16] "Optical Pressure Sensors | The Design Engineer's Guide | Avnet Abacus." [Online]. Available at https://www.avnet.com/wps/portal/abacus/ solutions/technologies/sensors/pressure-sensors/core-technologies/opti cal/ [accessed: 17-Jan-2020].

[17] Q. Yu and X. Zhou, "Pressure sensor based on the fiber-optic extrinsic Fabry-Perot interferometer," *Photonic Sens.*, vol. 1, no. 1, pp. 72–83, 2011.

[18] M. Han, "Theoretical and Experimental Study of Low-Finesse Extrinsic Fabry-Perot Interferometric Fiber Optic Sensors," *Diss. Theses*, pp. 1–142, 2006.

[19] F. Urban, J. Kadlec, R. Vlach, and R. Kuchta, "Design of a pressure sensor based on optical fiber Bragg grating lateral deformation," *Sensors*, vol. 10, no. 12, pp. 11212–11225, 2010.

[20] C. K. Harnett, H. Zhao, and R. F. Shepherd, "Stretchable optical fibers: Threads for strain-sensitive textiles," *Adv. Mater. Technol.*, vol. 2, no. 9, pp. 1–7, 2017.

[21] J. Guo, M. Niu, and C. Yang, "Highly flexible and stretchable optical strain sensing for human motion detection," *Optica*, vol. 4, no. 10, p. 1285, 2017.

[22] H. Zhao, K. O'Brien, S. Li, and R. F. Shepherd, "Optoelectronically innervated soft prosthetic hand via stretchable optical waveguides," *Sci. Robot.*, vol. 1, no. 1, pp. 1–10, 2016.

[23] I. M. Van Meerbeek, C. M. De Sa, and R. F. Shepherd, "Soft optoelectronic sensory foams with proprioception," *Sci. Robot.*, vol. 3, no. 24, p. eaau2489, 2018.

[24] P. A. Xu, A. K. Mishra, H. Bai, C. A. Aubin, L. Zullo, and R. F. Shepherd, "Optical lace for synthetic afferent neural networks," *Sci. Robot.*, vol. 4, no. 34, pp. 1–10, 2019.

[25] Z. Kappassov, D. Baimukashev, Z. Kuanyshuly, Y. Massalin, A. Urazbayev, and H. A. Varol, "Color-coded fiber-optic tactile sensor for an elastomeric robot skin," *Proc. – IEEE Int. Conf. Robot. Autom.*, vol. 2019-May, pp. 2146–2152, 2019.

[26] C. Lee et al., "Soft robot review," *Int. J. Control, Automation Syst.*, 2017, 15.1: 3–15.

[27] S. Yun et al., "Polymer-waveguide-based flexible tactile sensor array for dynamic response," *Adv. Mater.*, vol. 26, no. 26, pp. 4474–4480, 2014.

[28] A. Levi, M. Piovanelli, S. Furlan, B. Mazzolai, and L. Beccai, "Soft, transparent, electronic skin for distributed and multiple pressure sensing," *Sensors*, vol. 13, no. 5, pp. 6578–6604, 2013.

[29] Y. Kato, T. Mukai, T. Hayakawa, and T. Shibata, "Tactile sensor without wire and sensing element in the tactile region based on EIT method," in *SENSORS, 2007 IEEE*, 2007, pp. 792–795.

[30] D. O. Amoateng, M. Totaro, E. Falotico, and L. Beccai, "Intelligent Position, Pressure and Depth Sensing in a Soft Optical Waveguide Skin," In *2nd IEEE International Conference on Soft Robotics.*, no. 6, 2019, pp. 349–354.

[31] S. Zhao et al., "Programmable hydrogel ionic circuits for biologically matched electronic interfaces," *Adv. Mater.*, vol. 30, no. 25, p. 1800598, 2018.

[32] T. K. Bera and J. Nagaraju, "A labVIEW based data acquisition system for electrical impedance tomography (EIT)," In *Proceedings of the Third International Conference on Soft Computing for Problem Solving.* Springer, New Delhi, 2014. pp. 377–389.

[33] R. W. Stacey, "Electrical impedance tomography," Department of Energy by Department of Petroleum Engineering, Stanford University, 2006.

[34] M. Cheney, D. Isaacson, and J. C. Newell, "Electrical impedance tomography," *SIAM Rev.*, 1999, 41.1: 85–101.

[35] J. D. Kelleher, B. Mac Namee, and A. D'Arcy, *Fundamentals of Machine Learning for Predictive Data Analytics: Algorithms, Worked Examples, and Case Studies.* The MIT Press, Cambridge, 2015.

[36] J. Han, M. Kamber, and J. Pei, *Data Mining: Concepts and Techniques*, Elsevier, 2011.

[37] P. Cunningham, M. Cord, and S. J. Delany, "Supervised learning," In *Cognitive Technologies*, Springer, Berlin, Heidelberg, 2008. pp. 21–49.

[38] V. Podgorelec and M. Zorman, "Decision tree learning," In *Encyclopedia of Complexity and Systems Science*, 2015, 1–28.

[39] L. Rokach, "Ensemble-based classifiers," *Artif. Intel. Rev.*, 2010, 33.1–2: 1–39.

[40] N. Guenther and M. Schonlau, "Support vector machines," *Stata J.*, 2016, 16.4: 917–937.

[41] Faisal, M. F., Mohamed, A., Hussain, A., and Nizam, M. (2009). Support vector regression based S-transform for prediction of single and multiple power quality disturbances. *European Journal of Scientific Research*, 34(2), 237–251.

[42] G. I. Webb, "Lazy learning," In *Encyclopedia of Machine Learning and Data Mining*, Springer, 2017.

[43] S. Haykin, *Neural Networks: A Comprehensive Foundation*. Prentice Hall PTR, 1994.

[44] T. Borg, S. Cardona Mills, J. Scerri, C. Seguna, and K. Scicluna, "Integrated Position and Force Sensing for Optical Artificial Skin using Machine Learning Methods," In the *IEEE 45th Annual Conference of Industrial Electronics Society (IECON 2019), Lisbon, Portugal*, vol. 1, pp. 5401–5406, 2019.

[45] A. Majumdar, "Adaptive neuro-fuzzy systems in yarn modelling," In *Soft Computing in Textile Engineering*, 2010.

[46] M. P. Kazmierkowski, "Neural networks and fuzzy logic control in power electronics," In *Control in Power Electronics: Selected Problems*, 2003.

[47] M. Lo Preti, "Distributed Mechanical Sensing in a Soft Optical Skin," Unpublished master's thesis, https://etd.adm.unipi.it/t/etd-09222019-23 0248/, Pisa, Italy, 2019.

[48] M. Soni and R. Dahiya, "Soft eSkin: Distributed touch sensing with harmonized energy and computing," *Philos. Trans. A. Math. Phys. Eng. Sci.*, vol. 378, no. 2164, p. 20190156, 2020.

[49] T. G. Thuruthel, B. Shih, C. Laschi, and M. T. Tolley, "Soft robot perception using embedded soft sensors and recurrent neural networks," *Sci. Robot.*, vol. 4, no. 26, p. eaav1488, 2019.

[50] K. Chin, T. Hellebrekers, and C. Majidi, "Machine learning for soft robotic sensing and control," *Adv. Intel. Syst.*, 2020, 1900171.

5

Physical Contact Localization with Artificial Intelligence and Large-Area Fiber Bragg Grating Tactile Sensors for Collaborative Biorobotics

Tamás Czimmermann[1], Luca Massari[1], Jessica D'Abbraccio[1], Giuseppe Terruso[1], Martina Zaltieri[2], Giulia Fransvea[1], Andrea Aliperta[1], Eduardo Palermo[3], Emiliano Schena[2], Edoardo Sinibaldi[4] and Calogero Maria Oddo[1,*]

[1]The BioRobotics Institute of Scuola Superiore Sant'Anna and Department of Excellence in Robotics and AI of Scuola Superiore Sant'Anna, Pontedera, Italy
[2]Research Unit of Measurements and Biomedical Instrumentation, Center for Integrated Research, Universita' Campus Bio-Medico di Roma, Roma, Italy
[3]Center for Micro-BioRobotics, Istituto Italiano di Tecnologia, Pontedera, Italy
[4]Department of Mechanical and Aerospace Engineering Sapienza University of Rome, Rome, Italy
E-mail: tamas.czimmermann@santannapisa.it; luca.massari@santannapisa.it; jessica.dabbraccio@santannapisa.it; giuseppe.terruso@santannapisa.it; m.zaltieri@unicampus.it; giulia.fransvea@santannapisa.it; andrea.aliperta@santannapisa.it; eduardo.palermo@uniroma1.it; e.schena@unicampus.it; edoardo.sinibaldi@iit.it; calogero.oddo@santannapisa.it
*Corresponding Author

This chapter reports on the development of a novel large-area soft artificial skin with integrated Fiber Bragg Grating (FBG) sensors for the robotic perception of physical interaction. Specifically, by combining the FBG technology and machine learning techniques (fully connected neural network,

FCNN), the Presented artificial skin permits to simultaneously solve contact location and force, thus fostering applications in collaborative robotics.

5.1 Introduction

During the last decades, the interest in the field of robotics and the development of related technologies have led to a wider evolution of collaborative robotics, where humans and machines work in a close interaction sharing the same environment. A new generation of robots has been introduced to assist humans in performing several tasks, ranging from industrial production, transportation and delivery of goods, up to medical aid and rehabilitation. As interactions are becoming inevitably closer between robots and humans within these tasks, it is logic to equip them with a sort of tactile sensory experience. This is meant to increase robotic intelligence nevertheless permit safe and collaborative human–robot interactions, even in a highly dynamic scenario. Safe operation and interactions with their surroundings depend on the availability of sufficient tactile feedback [1], as it is in the … sensory modality that enables humans to explore objects' properties, develop body awareness, and interact with the surrounding environment through contact and manipulation.

In previous studies [2–4], the authors had designed and developed a polymeric artificial sensing skin, based on the FBG technology, with the aim to mimic the human sense of touch (Figure 5.1D). Such a technology allows localizing the point of contact and also measuring the applied force of a single indentation on the artificial skin. Also, a preliminary assessment of the performance of the tactile sensory system was carried out to evaluate the sensitivity. However, the localization and force prediction task based on the cross talk of adjacent sensors is an highly-challenging problem that is hard to model.

Deep learning is one of fastest growing fields in computer science, due to its ability to solve highly complex problems [5]. This ability of deep learning fits perfectly to solve the proposed challenge. We leveraged the FCNN technology to tackle the complex problem of localization and force detection.

5.2 Materials and Methods

This section is devoted to the description of the aforementioned system (Section 2.1), the experimental platform and datasets used to assess the

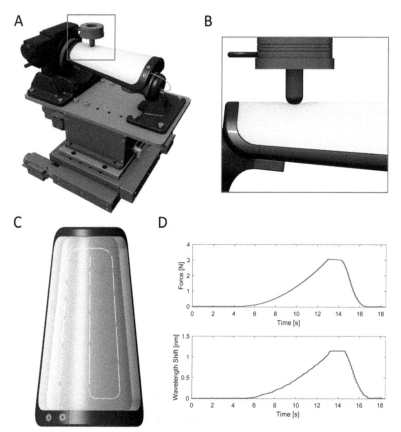

Figure 5.1 Experimental setup [2]. (**A**) Mechatronic platform for force-controlled indentation. (**B**) Inset of the indentation onto the polymeric matrix. (**C**) Embedded FGBs distributed in the polymeric skin. (**D**) Example of recorded data during an indentation above one of the embedded FBGs. The upper plot shows the exerted force as a function of time; the bottom plot shows the corresponding wavelength shift of the strained FBG.

capability of the artificial skin in perceiving the applied loads (Section 2.2), and the neural network structures evaluated during the experiments (Section 2.3).

5.2.1 FBG-based Sensing Skin

The authors employ a polymeric large-area artificial skin integrating FBG sensors to enable the perception of external force (localization and force detection of the applied pressure) in a three-dimensional (3D)-printed human-like forearm of a robotic manipulator. A single fiber can embed several FBGs to achieve significant sensing networks and high spatial resolution using tens (for a single fiber) or hundreds (when multiple fibers are simultaneously interrogated) of sensors. Each FBG resonates at λ_B: [6, 7].

$$\lambda_B = 2n_{eff}\Lambda.$$

where Λ is the spatial periodicity of the grating and n_{eff} the refractive index of the optical fiber. The application of interest needs the detection of force, therefore, an FBGs-based system is an excellent candidate, since the high sensitivity to strain is encoded in the λ_B shift. A typical sensitivity value is approximately 1 pm $\cdot$ $\mu\epsilon^{-1}$. Considering the high performance of the most common techniques for FBGs interrogation an FBG-based system can detect force causing strain around 1 pm [8]. In our application, the sensitivity, in terms of the relationship between the applied load and the occurred shift of the gratings Bragg wavelengths is shown in Figure 5.1(D) [2]. The proposed skin consists of an 8 mm thick polymeric layer (Dragon Skin 10 Medium; Smooth-on Inc, PA, USA) equipped with an 80 μm in diameter and 430 mm in length engineered optical fiber (FEMTOPlus Grating; FemtoFiberTec GmbH, Berlin, Germany). The optical fiber is endowed with 16 FBGs 8 mm in length, whose center-to-center distances and Bragg wavelengths (λ_B), were free with noncostant spacing. The optical fiber path – housing the 16 gratings – is inserted in the polymeric matrix to distribute the connected FBGs to cover the whole area. It is worth mentioning that the gratings are distributed in a custom configuration to allow cross talk between adjacent FBGs and therefore enable localization through measure of the strained FBGs (Figure 5.2). For the application of interest, optical fiber technology was used due to their small diameters that allowed low bending radius configurations. The silicone rubber allowed the fabrication of a thin, lightweight, and flexible substrate that offered interesting features, such as the capability to transmit the load to the buried FBGs, to stretch or wrinkle, preserving both the integrity of wires and sensors and the stability of the coverage. The coating covers a curved surface (15500 mm^2 in area). To survey the capabilities of the device, we combined it with deep learning strategies where we established different training/evaluation and test dataset, and we trained the machine learning system with the samples.

5.2.2 Experimental Platform and Datasets

To assess the capability of the developed artificial skin in perceiving position and intensity of the applied loads, a mechatronic platform was employed,

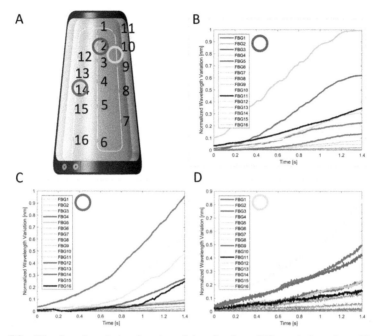

Figure 5.2 Wavelength variation function of time for three different indentations. The graph clearly shows the cross talk between adjacent FBGs.

[2] (Figure 5.1). This platform consisted of a 4 degrees of freedom system of motorized stages: a Cartesian manipulator made of a precision vertical (Z) positioner (8MVT120-25- 4247, STANDA, Vilnius, Lithuania) and X-Y stages (8MTF-102LS05), which guaranteed translational movements, while a rotator (8MR190-2) enabled 360° rotations. The apparatus enabled force-controlled indentations thanks to a 6-axis miniaturized load cell (Nano-43, ATI Industrial Automation, Apex, USA). Specifically, the load cell acted as a single axis sensor to perform the experiments in force-control mode of the platform along the loading direction of the skin (Z-axis). A customized probe (diameter of 11 mm, to mimic the size of a human fingertip), mechanically linked to the force sensor, was used to exert indentations across the skin's outer surface.

The FBGs required an interrogation unit able to illuminate the gratings with a broad spectrum, and detect the reflected wavelength (SmartScan, Smart Fibres Ltd, United Kingdom).

We established 2 experimental protocols in order to gather both training and test data to train FCNN (Figure 5.3).

A   B

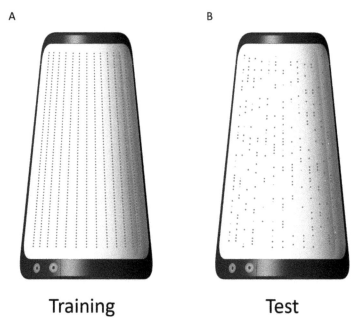

Training   Test

Figure 5.3   Locations of each indentation on the polymeric skin of **(A)** Training set; **(B)** Test set.

For training data, 1600 indentations were regulraly distributed on a 15500 mm^2 area of the artificial skin. Every indentation was applied with a force between 0 N and 4 N, and recorded at a sampling rate of 500 Hz. In the training dataset, all the training data had the single output value of each 16 FBGS sampled at a given time. All the training data were labeled by the current vertical (Y [mm]) and angular (R [°]) coordinates of the indentation and the currently applied force (F) measured by the load cell and sampled at the same given time as the FBGS output. The evaluation dataset was made of around 10% of the randomly selected samples of the training dataset. The total number of samples of the training dataset was 1713825, while the evaluation dataset had a size of 190425 samples.

For testing dataset, we performed 1600 indentations on random locations (Y, R) on the artificial skin with an applied force randomly selected in the range 0 N–4 N. The sampling and storing protocol of the dataset was the same as the training/evaluation dataset. The total number of samples in the testing dataset was 1904251.

An ad-hoc graphical user interface (GUI) was developed in LabVIEW (National Instruments, TX, USA) to control the instrumentation of the setup and enable data gathering and processing in real time. The deep learning system was developed in the nGene module of NI LabVIEW.

5.2.3 Neural Network Structures

At present, to determine the NN structure that suitably tackles a given problem requires some iterations. A key aspect to consider is to avoid overfitting. Namely, overfitting is a modeling error that occurs when a function is too closely fit to a limited set of data points [9]. Gradient descent is an iterative optimization algorithm used to minimize functions by moving toward the negative direction of the steepest descent. In machine learning, we use the gradient descent to update the parameters of our model. Parameters refer to coefficients and weights in neural networks [10]. Overfitting can be solved or the error can be reduced by some practices commonly used in deep learning field. The two most effective solutions are establishing and applying the correct dataset with the correct network structure.

Due to the aforementioned problems, more than 50 different network structures with different training parameters were evaluated during the experiments. In this study, we propose a selected one.

The main goal of the study was to train a neural network for regression, that means for a specific input it provides an output, ideally with minimal error from the desired output. For that, we established an FCNN, with a total of 579 hidden neurons in 9 hidden layers to calculate the value of 3 output neurons (namely for Y [mm], R [°], and F [N]), by fetching data into the 16 input neurons as the FBGs' sampled output. The structure of the network is depicted in Figure 5.4.

We applied rectified linear unit (ReLu) activation function in all of the hidden layers, except for the 5th, were a DropOut layer was employed. It is defined as $y = max(0, x)$. ReLU is linear (identity) for all positive values, and zero for all negative values, that means, it is fast to calculate as there is no complicated math [11]. The model therefore takes less time to train or run.

It also converges faster. Linearity means that the slope does not plateau, or "saturate," when x gets large. It does not present the vanishing gradient problem suffered by other activation functions such as *sigmoid* or *tanh*. ReLU is sparsely activated, and since it is zero for all negative inputs, it is likely for any given unit to not activate at all. These attributes tend to decrease overfitting and provides a better generalization of the networks.

The main function of the DropOut 5th layer is to prevent overfitting in the training database [12], by turning on and off every neuron in the layer by a

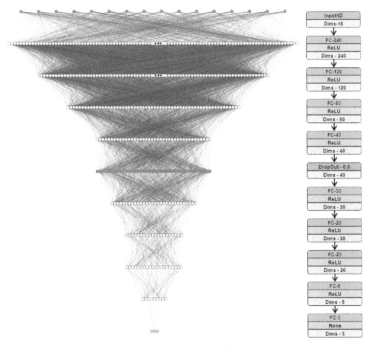

Figure 5.4   The structure and parameters of an FCNN employed for force prediction and localization. The output signal of the 16 FBGs provides the input for the network, that calculates three outputs, namely the y-coordinate, rotation angle, and force.

given chance. This can be considered as a natural random noise in the training samples that the network has to adapt to it during training. The adaptation also means a better generalization of the network, since every training iteration even on the same samples has a chance to be calculated differently, however, still gives the same result.

To train the network, we used the traditional way of back-propagation algorithm with a stochastic gradient descend optimizer. To optimize the training and drastically reduce the training time, we randomly grouped samples by 256 to create normalized mini-batches and performed the training on them, and then averaged the trained networks [13].

5.3 Results

The best performing networks were trained for 12 epochs on 1713825 training samples. The gradient of the loss function is shown in Figure 5.5. The training procedure was constantly monitored by performing a forward propagation

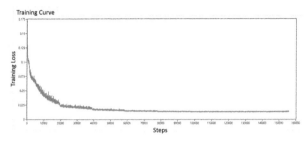

Figure 5.5   The curve of loss function after training.

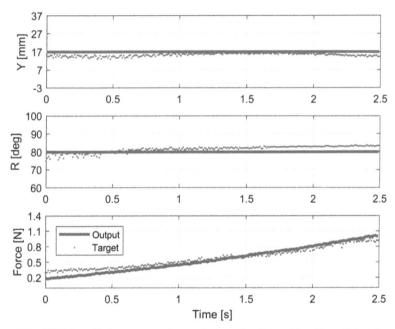

Figure 5.6   FCNN prediction vs target for one indentation for position (Y and R) and normal force (Fz). Errors: Y = 2.11 mm; R = 3.09° and F = 0.03 N.

and prediction in every 10th step using the evaluation dataset. The initial learning rate was set to 0.05 and was halved in every 1.5 step, with a weight decay of 10^{-8}.

We computed the results by performing prediction on the trained system by 1904251 test samples and averaged the errors by Y, R, and F. The system achieved an average of 5.51 mm error in Y, an 8.79° error in R, and average of 0.36 N prediction in F. Figure 5.6 depicts the errors of a single indentation performed during the recording of the test dataset.

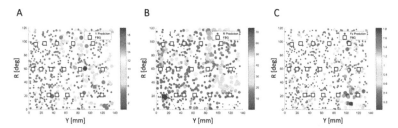

Figure 5.7   NN Prediction error of (**A**) Fz normal force (N), (**B**) R position (°), and (**C**) Y position (mm) for all the indentations. Higher the size of he circle higher the error, the color also shows the error (blue indicates little error, whereas red indicates big error). The white squares indicate the 16 FBGs.

The different panels of Figure 5.7 highlight the distribution of the errors among the whole set of indentations for the localization (Y [mm] and R [°]) and F [N]. The graph shows on the X-axis the rotation while on the Y-axis the vertical position (i.e. the coordinates of the indentations). The color and the size of the circles indicate the values of the errors.

5.4 Discussion and Conclusion

The work describes the development of a novel and useful subsystem for collaborative robotics. Namely, this study specifically exploits the hidden opportunities of a previously developed technology [2] to endow robots with the ability to sense tactile interactions.

As the individuals FBG tends to have a linear characteristics as output regarding the force of the indentation, grouping them up creates a very challenging problem of localization and force prediction based on their combined output. To achieve this goal, we equipped the artificial polymeric skin with 16 embedded FBGs with a trained, 11-layer large FCNN. To train the neural network, we established training/evaluation and test datasets by digitally sampling – with 500Hz – the wavelet variations of the embedded FBGs were over 1600 indentations.The system achieved an average of 5.51 mm error in Y, an 8.79° error in R, and average of 0.36 N prediction in force of the indentations.

As we visualize the errors of the predictions in Figure 5.7, we can state that the errors are mostly concentrating on certain areas of the polymeric skin. This spatial polarization mainly occurs due to the fabrication uncertainties/asymmetries of the artificial skin in these areas. The fabrication and implementation of the polymeric skin tend to have errors, more or less dense areas, and anchor points on skin's frame. These subfactors complicates the original problem, thus requiring a more sophisticated and complex feature extraction method from the machine learning algorithms. In future studies, we aim to overcome these feature-related problems by employing a more complex deep learning strategy that takes the physical characteristics of the polymeric skin into account.

Summarizing the conclusions, a novel application has been presented to provide tactile interaction sensation for collaborative robots. The study system performed with reasonably low error. However, there are still some questions regarding feature extraction. In the future, we aim to overcome these challenges to provide a more dynamic and ready-to-use solution for robotic close interaction.

References

[1] Y. Ohmura, Y. Kuniyoshi, and A. Nagakubo, "Conformable and scalable tactile sensor skin for curved surfaces," in *Proceedings 2006 IEEE International Conference on Robotics and Automation, 2006. ICRA 2006., IEEE, 2006*, pp. 1348–1353.

[2] J. D'Abbraccio, A. Aliperta, C. M. Oddo, M. Zaltieri, E. Palermo, L. Massari, G. Terruso, E. Sinibaldi, M. Kowalczyk, and E. Schena, "Design and development of large-area fbg-based sensing skin for collaborative robotics," in *2019 II Workshop on Metrology for Industry 4.0 and IoT (MetroInd4. 0&IoT)*, IEEE, 2019, pp. 410–413.

[3] P. Saccomandi, C. M. Oddo, L. Zollo, D. Formica, R. A. Romeo, C. Massaroni, M. A. Caponero, N. Vitiello, E. Guglielmelli, S. Silvestri, et al., "Feedforward neural network for force coding of an mri-compatible tactile sensor array based on fiber bragg grating," *Journal of Sensors*, vol. 2015, 2015.

[4] L. Massari, E. Schena, C. Massaroni, P. Saccomandi, A. Menciassi, E. Sinibaldi, and C. M. Oddo, "A machine-learning-based approach to solve both contact location and force in soft material tactile sensors," *Soft Robotics*, 2019.

[5] Z.-Q. Zhao, P. Zheng, S.-T. Xu, and X. Wu, "Object detection with deep learning: A review," *IEEE Transactions on Neural Networks and Learning Systems*, vol. 30, no. 11, pp. 3212–3232, 2019.

[6] T. Erdogan, "Fiber grating spectra," *Journal of Lightwave Technology*, vol. 15, no. 8, pp. 1277–1294, 1997.

[7] K. A. Othonos and G. E. Kohnke, "Fiber bragg gratings: Fundamentals and applications in telecommunications and sensing," *Physics Today*, vol. 53, no. 5, p. 61, 2000.

[8] D. Tosi, "Review and analysis of peak tracking techniques for fiber bragg grating sensors," *Sensors*, vol. 17, no. 10, p. 2368, 2017.

[9] D. M. Hawkins, "The problem of overfitting," *Journal of Chemical Information and Computer Sciences*, vol. 44, no. 1, pp. 1–12, 2004.

[10] L. Bottou, "Large-scale machine learning with stochastic gradient de- scent," in *Proceedings of COMPSTAT'2010*, Springer, 2010, pp. 177–186.

[11] P. Ramachandran, B. Zoph, and Q. V. Le, "Searching for activation functions," *arXiv preprint arXiv:1710.05941*, 2017.

[12] J. Ba and B. Frey, "Adaptive dropout for training deep neural networks," in *Advances in Neural Information Processing Systems*, 2013, pp. 3084–3092.

[13] M. Li, T. Zhang, Y. Chen, and A. J. Smola, "Efficient mini-batch training for stochastic optimization," in *Proceedings of the 20th ACM SIGKDD International Conference on Knowledge Discovery and Data Mining*, 2014, pp. 661–670.

6

Efficient Algorithms for Embedded Tactile Data Processing

Hamoud Younes[1,2,*], Mohamad Alameh[1], Ali Ibrahim[1,2], Mostafa Rizk[2] and Maurizio Valle[1]

[1]University of Genoa, Italy
[2]Lebanese International University, Lebanon
E-mail: hamoud.younes@edu.unige.it; mohammad.alameh@edu.unige.it;
ali.ibrahim@edu.unige.it; mostafa.rizk@liu.edu.lb; maurizio.valle@unige.it
*Corresponding Author

This chapter presents a survey of the existing algorithms and tasks applied for tactile data processing. The presented algorithms and tasks include machine learning, deep learning, feature extraction, and dimensionality reduction. Moreover, this chapter provides guidelines for selecting appropriate hardware platforms for the algorithm's implementation. The algorithms are compared in terms of computational complexity and hardware implementation requirements. A touch modality classification problem is addressed as a case study: FPGA implementations of two algorithms k-Nearest Neighbor (KNN) and Support Vector Machine (SVM) are detailed and analyzed. Both algorithms provided real-time classification consuming 236 mW and 1.14 W, respectively. Such results can be improved with the use of approximate computing techniques that provide a trade-off between performance and hardware resources usage. Speedups up to $2\times$ and $3.2\times$ along with 30% and 41% power reduction are obtained for KNN and SVM implementations, respectively.

6.1 Introduction

Electronic skin (e-skin) is being incorporated in a wide range of systems such as Internet of Things (IoT), robotics, industrial automation, and

113

prosthetics [1]. E-skin is composed of an array of tactile sensors, an interface electronics, an embedded processing unit (EPU), and a communication interface [2]. The EPU is responsible for: (1) extracting and processing information from raw sensory data and (2) supporting intelligent tasks such as classification or regression based on sophisticated and complex algorithms (e.g. machine- and deep learning). To accomplish these tasks, the EPU must fulfill a set of requirements in terms of computational complexity and implementation requirements (size, latency, and power consumption) depending on the target application.

The data acquired from tactile sensors corresponds to an electrical stimulus. The latter varies according to the type of the sensing material, dimensionality, responsiveness, and structure of the sensor. Processing algorithms employed by the EPU should be able to decode and efficiently handle the acquired data. Although traditional processing algorithms (e.g. Fourier transforms) and machine/deep learning algorithms are effective for tactile data processing, their use is bounded by computational complexity and hardware implementation performance [3–5]. For instance, a neural network is considered as an efficient solution for classification problems but implementing it on hardware platforms imposes several challenges such as low time latency, low power, etc. [6–8]. Three surveys addressed such challenges. Sze et al presented the challenges faced in the embedded systems in Ref. [9], and how circuit designers are to address these challenges in Ref. [10]. In Ref. [11], the authors presented an overview of the existing techniques that enable efficient implementations of machine/deep learning algorithms. This chapter shows the steps to be followed to choose a convenient hardware platform for the target application.

The main of contributions of this chapter can be summarized as follows: (1) it presents preprocessing algorithms to extract data from tactile sensors, (2) it offers a survey of classification and regression algorithms that can be embedded into e-skins, (3) it provides an algorithmic level computational complexity study and guidelines for targeting convenient hardware platform for the hardware implementation, and (4) it tackles a case study on touch modality classification.

The rest of the chapter is organized as follows: Section 6.2 provides an overview of the algorithms used for tactile data processing. Moreover, an assessment of the presented algorithms is provided in terms of classification accuracy and computational complexity. Section 6.3 presents the different hardware platforms that have been used for the implementation of these algorithms and the challenges faced. In Section 6.4, the FPGA implementation of two machine learning (ML) classifiers is presented as a case study for the embedded implementation of a touch modality classification. Finally, conclusions and future perspectives are highlighted in Section 6.5.

6.2 Tactile Data Processing Algorithms

Data processing algorithms presented in the literature could be divided into two categories: preprocessing and classification/regression. Preprocessing algorithms involve feature extraction and dimensionality reduction, while classification and regression algorithms are grouped into machine- and deep learning algorithms.

6.2.1 Data Preprocessing

Tactile data may be preprocessed to reduce noise and extract meaningful features. The extracted features could be (1) the variables that best describe the raw data and (2) the weights should be given for each variable. For instance, subsampling can be applied to a recorded touch reading to remove silent/noisy samples. Also, data obtained from certain taxels in the sensor patch can be considered in a pattern recognition problem. These taxels are the ones that provide reliable data (nonzero or unknown readings).

This section reports the algorithms presented in the literature for dimensionality reduction and feature extraction such as principal component analysis (PCA), independent component analysis (ICA), and linear discriminant analysis (LDA).

PCA is the base for multivariate data analysis (i.e. studying the effect of multiple variables on the output state) [12]. PCA is used for approximating data or reducing the dimensionality of the data e.g. representing data from X_n space in X_{n-k} space, where n and k are two positive integers. As a concrete example, if we have data with n features, then PCA helps to represent these data with $n - k$ features with the least possible losses. Figure 6.1 shows how PCA can be applied to reduce dimensionality from three dimensions (3D) to 2D (the figure has been generated using the data and code provided in Ref. [13]).

In Ref. [14], a finger-like shape tactile sensor has been used to collect data about fabric surfaces. Initially, fast Fourier transformation (FFT) was used to construct the original dataset, and then PCA was applied to compress the

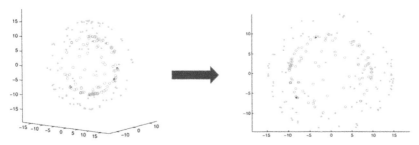

Figure 6.1   PCA example: 3D to 2D.

attribute data and extract feature information. In Ref. [15], kernel PCA [16] was used for low-resolution tactile image recognition for automated robotic assembly. Kernel PCA is a method to perform a nonlinear form of the PCA. It computes higher-order statistics among random variables while reducing the data dimensionality, thus being able to achieve the goal of both feature extraction and dimensionality reduction. Authors in Ref. [17] used local PCA [18] combined with a neural network to classify 16 household and toy objects. Local PCA is a nonlinear extension of the normal PCA. It has been used to obtain a less complex feature vector for the data obtained from tactile sensors mounted into a robotic arm.

ICA [19] can be seen as an extension of the PCA. It is a linear dimensionality reduction technique, which searches for the linear transformation that reduces or eliminates the linear dependency between elements of a random vector. An example of usage of ICA is the Cocktail Party Problem [20]. Spatial ICA has been adopted as a separation method that allows a robot to understand and interact with tactile information from multiple sources [21]. Figure 6.2 shows the procedure of tactile data separation from two objects using ICA along with time series clustering.

LDA shown in Figure 6.3 is yet another method for dimensionality reduction. It consists of finding the projection hyperplane that minimizes the variance within the same class, and maximizes the distance within the projected means of the classes [22].

Tactile images of deformable and nondeformable surfaces have been used for a classification problem in Ref. [23]. LDA has been used as a separation algorithm between six different surfaces with an accuracy rate of up to 95.5%. In Ref. [24], the authors have demonstrated the feasibility of using LDA for surface texture discrimination. Another use of LDA appears in Ref. [25] for terrain discrimination problems.

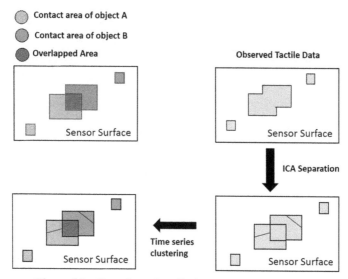

Figure 6.2 Procedure of tactile data separation using ICA.

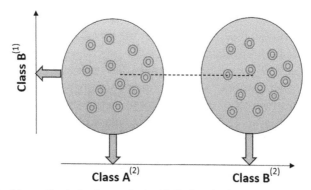

Figure 6.3 Linear discrimination analysis: (1) Bad projection and (2) Good projection.

6.2.2 Classification and Regression

6.2.2.1 Machine learning

Machine learning algorithms are an efficient solution for processing tactile data in various applications [26]. ML algorithms in general, can extract a complex, non linear input–output relationship based on learning by example approach. An ML algorithm is trained using a set of examples, where each example is described by a group of informative features. ML algorithms can support intelligent and predictive systems that can make accurate decisions on unseen data. ML algorithms are categorized into supervised and unsupervised algorithms. Supervised algorithms are aware of labeled input and output data while unsupervised algorithms are fed with unlabeled input data. Some of the ML algorithms used for tactile data processing belongs to the supervised learning category [25–28]. These algorithms include Naïve Bayes (NB), Decision Tree (DT), K-Nearest Neighbor (KNN), Support Vector Machine (SVM), Logitboost, etc.

NB is a probabilistic ML classifier that is based on Bayes' theorem given in Equation 6.1, with naïve independence assumptions between the features. It gives a quantitative approach to justify the evidence supporting a hypothesis i.e. the probability of occurrence of a certain action A given an action B has occurred. Generally, NB uses the Gaussian distribution parameters (e.g. mean, variance, etc.) on the dataset attributes for classification [29].

$$P(h/D) = \frac{P(D/h) \times P(h)}{P(D)} \qquad (6.1)$$

where $P(h)$ is the probability that a hypothesis h holds; $P(D|h)$ is the probability of observing data D given a hypothesis h; $P(D)$ is the probability that data D will be observed; and $P(h|D)$ is the probability that a hypothesis h holds given the observed data D.

Authors in Ref. [30] have used an NB classifier in an industrial application for a vegetable grading robot. A PIC32 microcontroller was used to obtain pressure data from two piezoresistive flexible tactile sensors mounted on a two-fingered robotic arm. The arm was able to classify green, moderate, and ripe vegetables with an accuracy of 85%. In Ref. [31], a Denso robotic arm equipped with embedded strain gauge and polyvinylidene fluorides (PVDF) in two layers on the finger is used to classify five different materials based on their surface texture. Using an NB classifier, the system achieved an accuracy of $73 \pm 10\%$.

Another supervised ML algorithm that is widely used is the DT. DTs reflect human-level thinking by exploring the simple logic behind data interpretations. In a DT approach, a feature is represented by a node, a decision is represented by a branch, and an outcome is represented by a leaf. Two main algorithms are suggested to be used for building the DT algorithm: (1) Classification and Regression Trees (CART) that uses Gini Index G (a quantity that measures the degree of inequality in the distribution of a given data) as a metric and (2) Iterative Dichotomiser 3 (ID3) that uses entropy function $H(S)$ and information gain $IG(A, S)$ as metrics [30]. The three metrics are given by the equations:

$$G = 1 - \sum_{t=0}^{t=k} P_t^2 \qquad (6.2)$$

where k is the possible class value and P_t is the probability of occurrence of class t.

$$H(S) = \sum_{c \in C} -p(c)\log_2 p(c) \qquad (6.3)$$

where S is the current dataset, C is the set of classes in S, and $p(c)$ is the proportion of the number of elements in C to the number of elements in S.

$$IG(A, S) = H(S) - \sum_{t \in T} p(t)H(t) \qquad (6.4)$$

where T represents the subsets created from splitting S by attribute A such that $S = \cup_{t \in T} t$.

An autonomous humanoid robot from Aldebaran Robotics called NAO has been equipped with a 1.6 GHz Intel Atom CPU and suited with an artificial skin that has a multimodal tactile sensor [27]. The NAO has been trained to recognize nine different touch modalities (e.g. scratch, tickle, rub, etc.). Recognition of up to 96.8% was obtained using the DT algorithm. The NAO has been also trained using KNN and SVM classifiers reaching a classification accuracy of up to 95.1% and 96.75%, respectively.

A frequently used ML algorithm for classification problems is the KNN. KNN assigns a class C for an unseen query point q based on the class of the K-nearest points to q from a training set S. The distance from q to $p_i \in S$ can be Chebyshev, Manhattan, or Euclidean distance. The latter is the most popular metric as given by Equation 6.5:

$$d_{q,p_i} = \sqrt{\sum_{i=0}^{i=N} (q_{F_i} - p_{F_i})^2} \qquad (6.5)$$

where F_i represents a feature of points q and p, and N is the size of the training set S.

Authors in Ref. [32] equipped an artificial fingertip with two perpendicular PVDF film sensors to acquire the surface roughness of eight standard solid nickel surfaces. Using a KNN classifier with $K = 9$, an average classification accuracy of $82.6 \pm 10\%$ has been attained. In Ref. [33], a tactile array sensor

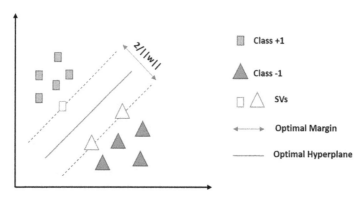

Figure 6.4   SVM hyperplane.

skin based on capacitive sensing technology is patched on the forearm of a
humanoid called "Cody." Cody is used to obtain the mobility and compliance
of 18 objects of different sizes. A KNN classifier with a variable value of
K has been used. The classifier has differentiated between four different
classes with a rate of 80% and $K = 2$. A higher classification accuracy of
up to 91.43% was obtained for a binary classification problem with $K = 4$.
Another binary classification problem for touch modality classification has
been studied in Ref. [34]. A KNN classifier was trained to recognize two
touch modalities: "rolling" and "sliding" recorded using a 4×4 tactile sensory
array. The K parameter has varied to 3,5 and 7, and 10-fold cross-validation
was applied to obtain credible results. The best classification accuracy was
89.6% for $K = 3$.

Another robust discriminative algorithm is SVM. SVM classification
embraces the concept of a decision boundary that separates two different
classes. This boundary is in the form of a hyperplane that is constructed in
the training phase. The data points that lie on the boundary lines are called
support vectors (SVs) as shown in Figure 6.4. These SVs are required for the
classification phase.

Consider a training data labeled as $(x_i, y_i), i = 1, 2, \ldots, N$, $y_i \in$
$\{-1, +1\}$ and $x_i \in R^d$. The hyperplane is the plane that separates the two
classes of squares and triangles. Any pattern x that belongs to the hyperplane
in the feature space can be described by Equation 6.6, where w is a normal
vector to the hyperplane and b is a constant

$$wx + b = 0 \tag{6.6}$$

The data points are separated by the two hyperplanes described in Equations 6.7 and 6.8, thus, the main objective is to maximize the distance between them:

$$wx + b = +1 \tag{6.7}$$
$$wx + b = -1 \tag{6.8}$$

The optimum separation hyperplane conditions can be formulated into the expression given by Equation 6.9 where the distance $\| w \|$ needs to be minimized:

$$y_i(wx_i + b) \geq 1, i = 1, \ldots, N \tag{6.9}$$

In most ML classification problems, the training data can't be linearly separated in the original space. Thus, the input space is mapped to a higher dimensional one where linear separation is feasible. Such mapping is a computationally expensive task, especially for large-scale applications. Therefore, SVMs utilize kernel functions $K(x_i, x_j)$ that replace the inner product of the optimization problem in Equation 6.9 as given in Equation 6.10:

$$y_i(K(x_i, x_j) + b) \geq 1, i = 1, \ldots, N \tag{6.10}$$

The most common kernel functions used are linear, polynomial, sigmoid, and Gaussian radial basis function (RBF) [35]. In the SVM classification phase, a new unseen sample is classified according to the function given in Equation 6.11:

$$F(x) = \text{sign}\left(\sum \alpha_i y_i K(x_i, x) + b\right) \tag{6.11}$$

Five objects that have the same size, but different weights are covered with the most common surface textures (e.g. rough, sand, glass, etc.). These objects have been used to test the ability of an NAO robot to classify objects using an SVM classifier [28]. The paper claims a classification accuracy of up to 100%. In Ref. [36], a robotic hand with 5 fingers and 20 active degrees of freedom was equipped with a BioTac sensor to classify 20 daily used objects (e.g. ball, bottle, sponge, etc.). A least square SVM (LS-SVM) [37] classifier was adopted and a discrimination accuracy rate up to 97% was achieved. SVM has been also used in Ref. [32] for differentiating between eight nickel surfaces based on their roughness and it provided an accuracy rate up to 78.8 $\pm$ 14% using an RBF kernel.

Another set of ML algorithms that has been used for tactile data processing and reported in the literature includes locally weighted projection

Table 6.1   Computational complexity of machine learning algorithms

Algorithms	Applications	Training	Classification
Naïve Bayes	Classification	$O(nf)$	$O(f)$
Decision Tree	Classification/Regression	$O(n^2f)$	$O(f)$
SVM (kernel based)	Classification/Regression	$O(n^2f + n^3)$	$O(n_{SV}f)$
KNN	Classification/Regression	–	$O(nf)$
Linear Regression	Regression	$O(f^2n + f^3)$	$O(f)$
Random Forest	Classification/Regression	$O(n^2fn_{trees})$	$O(fn_{trees})$

regression [38], extreme learning machine [39], regularized extreme learning machine [40], and K-means clustering [41]. All these algorithms achieved a classification accuracy rate of $85 \pm 10\%$.

These algorithms will be implemented on different hardware devices/platforms to be used as an EPU for the designed e-skin. When embedding ML algorithms in the e-skin, the implementation must maintain the hardware complexity, latency, and energy consumption as low as possible for portable and battery-powered devices. This means that an algorithm might support a certain application with high accuracy but contradicts with the available hardware space. Thus, a trade-off between the required accuracy and the available hardware space is to be considered. The trade-off is directly related to the computational complexity of these ML algorithms. Table 6.1 shows the computational complexity of the most commonly used algorithms in the Big-O notation [42], where n is the size of the training set, f is the number of features, n_{trees} is the number of trees, and n_{SV} is the number of support vectors.

The complexity given in Table 6.1 has been analyzed based on the degree of complexity provided by Figure 6.5. It is noticed that algorithms such as DT and SVM involve complex training phase that increases quadratically for a large number of training points. For linear regression (LR), the training phase complexity also increases quadratically with the number of features, which is usually less than the number of training points. Meanwhile, the DT, SVM, and LR classification phases are relatively less complex. For NB, the training phase is less complex compared to the SVM and DT with a low complexity classification phase too. This is due to the linear complexity compared to the quadratic one in the case of the SVM and DT. Similarly, the complexity of the classification phase of the KNN increases linearly with the increase in the number of training points and the number of features, and the same is

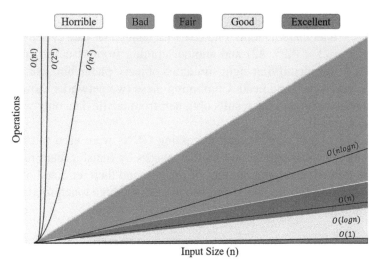

Figure 6.5 Big-O complexity.

observed from Equation 6.5. Although KNN doesn't have a separate training phase, it imposes a higher complexity compared to algorithms such as SVM ($n \gg nSV$).

6.2.2.2 Deep learning

Deep learning (DL) is a kind of artificial neural network (NN) where the network has more hidden layers inside it [43]. Usually, DL is used as a classification and feature extraction method at the same time (especially in image processing); so, no handcrafting of features is required [44, 45].

A convolutional neural network (CNN) is a deep neural network – the name is derived from the convolutional layers used in this network – that is usually used in image processing. In tactile sensing, different works used DNN or CNN for tactile data classification.

In Ref. [7], the CNN is used to evaluate four attributes for 23 gel-like foods: elasticity, smoothness, stickiness, and granularity. Four separate CNNs were trained where the input is a time series image coming from a pressure sensor. The image is 44×44 pixels (without the boundaries) in size. The network has four outputs, and each one corresponds to the sensory evaluation of a single attribute e.g. elasticity.

A high-resolution tactile sensor attached to a robotic arm was used to collect pressure maps of 22 daily-life objects, of 28×50 pixels each in Ref. [46]. Different CNNs were used, some pretrained on millions of images

[47], and others were built from the scratch. A combination of two CNNs was used for tactile data classification: one CNN for sequential data coming from tactile sensors $(32 \times 32 \times 32)$ and another coming from visual interesting points (25×3) for identifying eight simulated objects: plane, bird, car, chair, hand, vase, quadruped, and head. Combining these two networks showed an improvement compared to the results obtained from tactile data only without visual guidance [48].

In Ref. [49], benchmark image processing CNNs were used to classify tensorial data collected from 4×4 tactile sensors by transfer learning. The CNNs were trained on a large number of images, and then retrained on synthetic images coming from tactile sensors to classify three touch modalities: brushing, rolling, and sliding.

6.3 Embedded Processing System

6.3.1 Hardware Platforms

The hardware platform must be able to handle the complexity of the algorithm while achieving the expected performance in terms of time latency and energy consumption. A wide selection of hardware devices and platforms maybe used to implement the tactile processing algorithms. Some sound and widely used devices include field programmable gate array (FPGA), graphics processing unit (GPU), microcontroller unit (MCU), parallel ultralow-power platform (PULP), tensor processing unit (TPU), application-specific integrated circuit (ASIC), and platforms include Raspberry Pie, ZedBoard, Zynqberry, Python Productivity for Zynq (PYNQ), etc.

The available hardware devices and platforms differ in size, target programming language, area utilization (LUT, FF, DSP, BRAM, etc.), maximum operating frequency, etc. Table 6.2 presents the common characteristics of the most used hardware devices and platforms related to the variety of machine- and deep learning applications.

Taking into consideration the information presented in Table 6.2, an FPGA is suitable for implementing simple/moderate ML algorithms such as the LR or DT, while a neural network is the best fit on a GPU. Similarly, a more complex algorithm such as the KNN/SVM can be implemented on the FPGA, PULP, or ZedBoard/Zynqberry but the best choice depends on the application requirements. For example, choosing the GPU for a wearable device (e.g. smart watch) is not feasible, and the same can be said for implementing AI training on a microcontroller. Besides, the development

Table 6.2 Comparison of hardware devices/platforms

Types	Names	Framework/ Programming Languages	Strengths	Weaknesses
Device	FPGA	VHDL, Verilog, C/C++ with OpenCL, SDAccel, HLS	High performance per watt, parallelism	Not suited for floating-point operations, long development time, programming difficulty
	GPU	OpenCL, NVIDIA CUDA, C/C++, Java, Python	Massive processing power for image, video, and signal processing	High power consumption, need for API frameworks to take advantage of parallelism
	ASIC	Application-specific, ex: TensorFlow for TPUs, tools from manufactures	Optimum combination of performance and power consumption	High cost, long development time, not configurable
Platform	PULP	C language only	Low power consumption, tunable performance, Open source	Low size on-chip memory, long development time
	ZedBoard/Zynqberry	– adopts characteristics of FPGA and ARM processors – Pluses: Ability to use FPGA as a hardware accelerator, Linux Development		
	PYNQ	– adopts characteristics of FPGA and ARM processors – Pluses: Python Programming, Arduino and Raspberry Pie shield connectors		

time is a key issue to consider, especially for complex ML/DL algorithms such as implementing a CNN using VHDL language. All these considerations are to be discussed before selecting an appropriate hardware platform.

In general, the GPU is one of the best performers for fast ML processing. This is due to the available high memory bandwidth and a large number of processing cores. On the other hand, GPU computations can be about four times more expensive than CPU computations. So, if the gain in speedup is

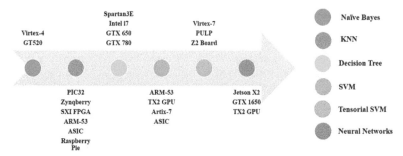

Figure 6.6 Hardware platforms used for different ML and deep learning algorithms.

not relative to that cost, CPUs can be more suitable than GPUs. Moreover, in each ML application, the training and testing implementations are not correlated. Hence, the GPU can be used for neural network training, while preserving the CPU for testing purposes. The same analogy can be viewed for all the hardware platforms. Figure 6.6 shows the hardware platforms reported in the literature that are used with the increased complexity of ML algorithms.

Although Section 6.2 provided an overview of machine- and deep learning algorithms that have been used for tactile data processing, the implementation of these algorithms on hardware platforms is still a challenge. The hardware implementations of the above-highlighted algorithms will be assessed even for different applications i.e. not limited to tactile data. This is because of the lack of such implementations for tactile data processing. Moreover, this assessment will help in studying the feasibility of implementing the algorithm itself on each hardware platform.

Table 6.3 presents some the hardware implementations of different ML algorithms. The implementation of Naïve Bayes has been carried out on Virtex-4 FPGA consuming a total of 2% occupied area for handwriting recognition problem of 70,000 samples [50]. In Ref. [51], a parallel architecture of NB has been implemented using the GPUs. For document classification problem, a speedup up to $34\times$ and $11\times$ can be achieved compared to sequential and parallel versions, respectively, using CPU. The GPU used was GeForce GT520 2GB graphics card and the number of used documents is 861,454. A pipelined DT implementation on FPGA has been presented in Ref. [52]. The development board used was Digilent Nexys-2 Spartan-3E FPGA board. The complete DT model required 6442 LUTs, 5336 FFs, and 22 block RAMs, which resemble 62% of area utilization. For a tree with

Table 6.3 Embedded machine learning on different hardware devices/platforms

Algorithms	Devices/Platforms	Implementation Results
Naïve Bayes	Virtex-4 FPGA	2% area utilization
	GeForce GT520 GPU	34× speedup compared to Intel Core 17-2600 CPU operating at 3,40 GHz,
Decision Tree	Digilent Nexys-2 Spartan-3E FPGA	62% area utilization Time latency of 220 ns for a tree with 100 tuples
	GeForce GTX 650 Ti	485× speedup compared to quadcore processor (Intel Core i7-870, 8M Cache, 2.93 GHz),
SVM	ARM 53 processor	1530 mW power consumption running at 990 MHz
	Jetson TX2 GPU	29× speedup compared to ARM 53 Processor 2090 mW power consumption running at 854 MHz
	Artix-7 FPGA	6× increase in throughput compared to Raspberry Pi 3B 1/5 power reduction compared to Raspberry Pi 3B
	Virtex-7 FPGA	30% area utilization A peak performance of 302 G-ops while consuming 1.14 W
	PULP	34x speedup compared to ARM Cortex M4 running at 168 MHz Power consumption <150 mW
	ASIC	0.3 mm^2 area utilization using 65 nm CMOS technology Energy of 13.4 nJ running at 5 Hz nominal frequency 42x energy efficient more than Artix-7 FPGA
KNN	ASIC	0.16 mm^2 area utilization using 65 nm CMOS technology Energy of 0.31 nJ running at 59 Hz nominal frequency 12x energy efficient more than Artix-7 FPGA
	Artix-7 FPGA	5x increase in throughput compared to Raspberry Pi 3B 1/4 power reduction compared to Raspberry Pi 3B
	ARM 53 processor	Power consumption of 1480 mW running at 990 MHz
	Jetson TX2 GPU	29x speedup compared to ARM 53 Processor Power consumption of 2120 mW running at 854 MHz

100 tuples, a time latency of 220 ns was recorded. A set of NVIDIA boards was used to evaluate the performance of DT models for large-scale data [53]. A maximum mean speedup of $585\times$ for a dataset of 10 M instances was obtained. The authors in Ref. [54] have implemented the KNN and SVM classifiers on ARM 53 processors, Jetson TX2 GPU, Artix-7 FPGA, and ASIC. The proposed SVM and KNN classifiers on the ASIC platform occupy an area of 0.17 mm^2 and 0.3 mm^2 while dissipating 39.4 mW and 76.9 mW, respectively. The experimental results showed that the use of FPGA and ASIC lead to the highest throughput (decision/s) as well as the lowest power consumption. The obtained results were also superior to the one obtained when using Raspberry Pi.

For a touch modality classification problem, two architectures for tensorial SVM [2] and DCNN [49] have been applied on Virtex-7 FPGA and Jetson TX2 development board from NVIDIA, respectively. The FPGA implementation provided a real-time classification and a power consumption of 1.14 W. In Ref. [55], a tactile data decoding module using SVM based tensor kernel algorithm for touch modalities was implemented on PULP [56]. The decoding module ensured a power consumption of less than 150 mW for a wearable device requirement. The proposed implementation runs $34\times$ faster than an ARM Cortex M4 running at 168 MHz at the same power consumption. In Ref. [57], an Enclustra SX1 FPGA was used to process the tactile data obtained by a novel Hex-O-Skin, while a PIC32 MCU has been utilized to obtain the pressure data from two PVDF tactile sensors [30]. Another 32-bit microcontroller, Teensy 3.2, was adopted as the main unit of a Vibrotactile Stimulation system [58].

6.4 Case Study: Touch Modality Classification

The touch modality classification problem has been the focus of several works in the literature e.g. [26, 27, 39]. We have surveyed the literature for the different ML algorithms used for touch modality classification. Based on this survey, most of the works reported that the SVM and KNN are the most effective ML algorithms to deal with this problem.

This section introduces the experimental setup used for touch modality classification in terms of the used dataset, preprocessing techniques and the performance of the algorithms in terms of classification accuracy. Then, the FPGA implementation of these two algorithms is presented and analyzed.

6.4.1 Experimental Setup

The touch modality problem we are targeting is the binary classification problem i.e. "sliding a finger" vs "rolling a washer" [26]. The dataset used contains data for 70 participants. Each participant performed a touch on a 4×4 tactile sensor for 10 seconds on both the horizontal and vertical directions. Thus, the final dataset contained 280 patterns. Using a 3 kHz sampling frequency, each touch was presented as a tensor of size $4 \times 4 \times 30,000$, where 30,000 raw samples were recorded from the 4×4 sensor during the 10 s duration. KNN [34] and SVM [26] classifiers based on the tensorial representation of input were recently proposed for the binary classification problem.

- KNN classifier: The authors have applied a feature extraction process on the initial dataset. First, the samples outside the range of 3.5 to 7 s were removed as they involved static movement or noisy information. Then, the mean of each 30,000 samples was calculated, resulting in a tensor of size $4 \times 4 \times 1$. Several simulation scenarios were studied and reported in Table 6.4 [34].
- SVM classifier: The authors have considered the data acquired in the first 7 out of 10 s using the same sampling frequency of 3 kHz. This resulted in a tensor size of $4 \times 4 \times 21000$. Such tensor size imposed an impractical computational task. Thus, the amount of energy provided by each single element of the sensor was analyzed. This task showed that only a portion of the 21,000 elements carry actual information. Then, a subsampling strategy was applied to find the best tensor 3rd dimension size D. The different simulations carried out are summarized in Table 6.5 [26] where λ, σ, and α represent the kernel parameters used.

The results in Tables 6.4 and 6.5 justify the use of KNN and SVM to support the touch modality classification problem with their classification accuracy. Recalling that these ML algorithms will be embedded into an electronic skin, the following section details the implementation of these

Table 6.4   KNN classification results

Training Dataset Size	Classification Accuracy (%)		
	$K = 3$	$K = 5$	$K = 7$
80% split	84	80.3	79
85% split	86	83.3	81
90% split	82.1	82.1	82.1
10-fold Cross Validation	89.6	89.3	89

Table 6.5   SVM classification results

Simulation	Classification Accuracy (%)		
Scenario	$D = 20$	$D = 50$	$D = 100$
Run #1	85 ($\lambda = 0.1$, $\sigma = 1$, $\alpha = Qz/2$)	83.5 ($\lambda = 0.1$, $\sigma = 1$, $\alpha = Qz/2$)	83.5 ($\lambda = 0.1$, $\sigma = 1$, $\alpha = Qz/2$)
Run #2	87.5 ($\lambda = 1$, $\sigma = 2^4$, $\alpha = Qz/2$)	85 ($\lambda = 10$, $\sigma = 2^3$, $\alpha = Qz/2$)	90 ($\lambda = 1$, $\sigma = 2^{-1}$, $\alpha = 0$)
Run #3	80 ($\lambda = 0.1$, $\sigma = 2^1$, $\alpha = Qz/2$)	87.5 ($\lambda = 1$, $\sigma = 2^2$, $\alpha = Qz/2$)	90 ($\lambda=1$, $\sigma=2^2$, $\alpha= Qz/2$)

algorithms on FPGA by exploring the hardware area occupied, time latency, and power consumption.

6.4.2 Implementation Details

The FPGA implementation of the tensorial SVM classifier is reported in Ref. [2]. The paper proposed two different architectures: cascaded and parallel to reach an adequate trade-off between real-time functionality and hardware resources. The parallel implementation reported 1.14 W power consumption, while achieving a peak performance of 302 G-ops. A tensor size of $8 \times 8 \times 20$ is used to represent the input data.

As for the KNN, the implementation is carried out using high-level synthesis (HLS) on Zynqberry [59]. The KNN classifies the unseen sample by executing the following steps: (1) distance calculation from the unseen sample to all the training samples, (2) the calculated distances are sorted in ascending order, and (3) the $K = 3$ neighbours with the smallest distances are chosen, and the output class is the class of the majority of the three neighbors. The KNN classifier was coded in C++ and optimized using Vivado HLS directives. Then, it was exported as an RTL intellectual property (IP) block. The IP was imported into Vivado to obtain the implementation report. The report showed that the KNN classifier consumes 236 mW while classifying a new sample within 1 ms.

Tables 6.6 and 6.7 summarize the outcome of the implementations on FPGA. KNN was implemented on the Zynqberry Platform (XC7Z010ICLG225-1L FPGA) operating at 100 MHz. SVM implementation was carried out on Virtex-7 XC7VX980T operating at 120 MHz.

The obtained results illustrate the feasibility of the implementation of the most used ML algorithms for tactile data processing. For embedding these algorithms into e-skin, some observations must be considered. For KNN,

Table 6.6 Implementation details for SVM on FPGA

Algorithm	Tensor Size	Training Tensors	Occupied Area	Time Latency	Power Consumption
SVM	$8 \times 8 \times 20$	100	13%	<200 ms	1.14 W

Table 6.7 Implementation details for KNN on FPGA

Algorithm	Tensor Size	Training Tensors	Occupied Area	Time Latency	Power Consumption
KNN	$4 \times 4 \times 1$	280	3%	<3 ms	236 mW

although the obtained power consumption is low, this could increase dramatically if the training size is large. As for SVM, the power consumption is relatively high, and the occupied area may not be suitable for e-skin size. It is worth noting that these two algorithms use a tensor-based input data, which is reported to preserve the initial information of the touch [60].

Taking these results into consideration with the complexity study in Table 6.1, and the implementations reported in the literature, several solutions were proposed to decrease the complexity of the embedded ML implementations [54, 55]. One of the effective solutions is the use of approximate computing techniques (ACTs). The authors in Ref. [61] have presented an approach for applying algorithmic level ACTs for the discussed KNN and SVM implementations. A circuit-level ACT is presented in Ref. [62] that utilizes the use of inexact accumulators for ML algorithms. For instance, the classification phase of the KNN and SVM can be accelerated up to $2\times$ and $3.2\times$ while achieving 30% and 41% power reductions, respectively, when applying algorithmic-level ACTs. Also, power savings up to 69% is achieved when the inexact accumulators are used in the SVM classifier implementation.

6.5 Conclusion

This chapter introduced the state-of-the-art processing algorithms used for tactile data processing. Starting with feature extraction algorithms to obtain a meaningful representation of data, the chapter proceeds to machine- and deep learning algorithms used for classification and regression applications. Then, several existing hardware platforms were studied as candidates for embedding the algorithms in the e-skin. Finally, two of the presented algorithms (KNN and SVM) were adopted for a touch modality classification problem. Similarly, the implementation of these algorithms was conducted on two different platforms: Virtex-7 and Zynqberry. The obtained results showed a low latency and area size while delivering considerably challenging power consumption in the case of the SVM. Such a challenge has been tackled using approximate computing, reaching a speedup of up to $3.2\times$ with 41% power reductions, without affecting the quality of the target application.

References

[1] L. Seminara et al., 'Towards integrating intelligence in electronic skin', *Mechatronics*, vol. 34, pp. 84–94, Mar. 2016, doi: 10.1016/j.mechatronics.2015.04.001.

[2] A. Ibrahim and M. Valle, 'Real-time embedded machine learning for tensorial tactile data processing', *IEEE Trans. Circuits Syst. Regul. Pap.*, vol. 65, no. 11, pp. 3897–3906, Nov. 2018, doi: 10.1109/TCSI.2018.2852260.

[3] J. Qiu, Q. Wu, G. Ding, Y. Xu, and S. Feng, 'Erratum to: A survey of machine learning for big data processing', *EURASIP J. Adv. Signal Process.*, vol. 2016, no. 1, p. 85, Dec. 2016, doi: 10.1186/s13634-016-0382-7.

[4] U. S. Shanthamallu, A. Spanias, C. Tepedelenlioglu, and M. Stanley, 'A brief survey of machine learning methods and their sensor and IoT applications', in *2017 8th International Conference on Information, Intelligence, Systems & Applications (IISA)*, Larnaca, Aug. 2017, pp. 1–8, doi: 10.1109/IISA.2017.8316459.

[5] S. Pouyanfar et al., 'A survey on deep learning: Algorithms, techniques, and applications', *ACM Comput. Surv.*, vol. 51, no. 5, pp. 1–36, Jan. 2019, doi: 10.1145/3234150.

[6] Z. Hajduk, 'Reconfigurable FPGA implementation of neural networks', *Neurocomputing*, vol. 308, pp. 227–234, Sep. 2018, doi: 10.1016/j.neucom.2018.04.077.

[7] A. Shibata, A. Ikegami, M. Nakauma, and M. Higashimori, 'Convolutional neural network-based estimation of gel-like food texture by a robotic sensing system', *Robotics*, vol. 6, no. 4, p. 37, 2017.

[8] C. Zhang, P. Li, G. Sun, Y. Guan, B. Xiao, and J. Cong, 'Optimizing FPGA-based accelerator design for deep convolutional neural networks', in *Proceedings of the 2015 ACM/SIGDA International Symposium on Field-Programmable Gate Arrays – FPGA'15*, Monterey, California, USA, 2015, pp. 161–170, doi: 10.1145/2684746.2689060.

[9] V. Sze, Y.-H. Chen, J. Emer, A. Suleiman, and Z. Zhang, 'Hardware for machine learning: Challenges and opportunities', in *2017 IEEE Custom Integrated Circuits Conference (CICC)*, Austin, TX, Apr. 2017, pp. 1–8, doi: 10.1109/CICC.2017.7993626.

[10] V. Sze, 'Designing hardware for machine learning: The important role played by circuit designers', *IEEE Solid-State Circuits Mag.*, vol. 9, no. 4, pp. 46–54, 2017, doi: 10.1109/MSSC.2017.2745798.

[11] M. Osta, M. Alameh, H. Younes, A. Ibrahim, and M. Valle, 'Energy efficient implementation of machine learning algorithms on hardware platforms', presented at the *26th IEEE International Conference on Electronics Circuits and Systems*, pp. 21–24, Genova, Italy, Nov. 2019.

[12] S. Wold, K. Esbensen, and P. Geladi, 'Principal component analysis', *Chemom. Intell. Lab. Syst.*, vol. 2, no. 1–3, pp. 37–52, 1987.

[13] L. Derksen, 'Visualising high-dimensional datasets using PCA and t-SNE in Python', *Medium*, Apr. 29, 2019. https://towardsdatascien ce.com/visualising-high-dimensional-datasets-using-pca-and-t-sne-in-python-8ef87e7915b (accessed June 02, 2020).

[14] H. Hu, Y. Han, A. Song, S. Chen, C. Wang, and Z. Wang, 'A finger-shaped tactile sensor for fabric surfaces evaluation by 2-Dimensional active sliding touch', *Sensors*, vol. 14, no. 3, pp. 4899–4913, Mar. 2014, doi: 10.3390/s140304899.

[15] Y.-H. Liu, Y.-T. Hsiao, W.-T. Cheng, Y.-C. Liu, and J.-Y. Su, 'Low-resolution tactile image recognition for automated robotic assembly using kernel PCA-based feature fusion and multiple kernel learning-based support vector machine', *Math. Probl. Eng.*, vol. 2014, pp. 1–11, 2014, doi: 10.1155/2014/497275.

[16] B. Schölkopf, A. Smola, and K.-R. Müller, 'Kernel principal component analysis', in *International Conference on Artificial Neural Networks*, Lausanne, Switzerland, Oct. 1997, pp. 583–588.

[17] M. Schopfer, H. Ritter, and G. Heidemann, 'Acquisition and application of a tactile database', in *Proceedings of the 2007 IEEE International Conference on Robotics and Automation*, Apr. 2007, pp. 1517–1522, doi: 10.1109/ROBOT.2007.363539.

[18] A. Weingessel and K. Hornik, 'Local PCA algorithms', *IEEE Trans. Neural Netw.*, vol. 11, no. 6, pp. 1242–1250, Nov. 2000, doi: 10.1109/72.883408.

[19] P. Comon, 'Independent component analysis. A new concept?', *Signal Process.*, vol. 36, no. 3, pp. 287–314, Apr. 1994, doi: 10.1016/0165-1684(94)90029-9.

[20] S. Haykin and Z. Chen, 'The cocktail party problem', *Neural Comput.*, vol. 17, no. 9, pp. 1875–1902, 2005.

[21] K. Lee, T. Ikeda, T. Miyashita, H. Ishiguro, and N. Hagita, 'Separation of tactile information from multiple sources based on spatial ICA and time series clustering', in *2011 IEEE/SICE International Symposium on System Integration (SII)*, Kyoto, Japan, Dec. 2011, pp. 791–796, doi: 10.1109/SII.2011.6147549.

[22] P. Xanthopoulos, P. M. Pardalos, and T. B. Trafalis, 'Linear discriminant analysis', in *Robust Data Mining*, P. Xanthopoulos, P. M. Pardalos, and T. B. Trafalis, Eds. New York, NY: Springer, 2013, pp. 27–33.

[23] M. Pal, A. Khasnobish, A. Konar, D. N. Tibarewala, and R. Janarthanan, 'Classification of deformable and non-deformable surfaces by tactile image analysis', in *Proceedings of The 2014 International Conference on Control, Instrumentation, Energy and Communication (CIEC)*, Calcutta, India, Jan. 2014, pp. 626–630, doi: 10.1109/CIEC.2014.6959165.

[24] H. Nguyen et al., 'Dynamic texture decoding using a neuromorphic multilayer tactile sensor', in *2018 IEEE Biomedical Circuits and Systems Conference (BioCAS)*, Cleveland, OH, Oct. 2018, pp. 1–4, doi: 10.1109/BIOCAS.2018.8584826.

[25] W. Kalas, 'Tactile sensing for ground classification', *J. Autom. Mob. Robot. Intell. Syst.*, vol. 7, no. 2, pp. 18–23, 2013.

[26] P. Gastaldo, L. Pinna, L. Seminara, M. Valle, and R. Zunino, 'Computational intelligence techniques for tactile sensing systems', *Sensors*, vol. 14, no. 6, pp. 10952–10976, June 2014, doi: 10.3390/s140610952.

[27] M. Kaboli, A. Long, and G. Cheng, 'Humanoids learn touch modalities identification via multi-modal robotic skin and robust tactile descriptors', *Adv. Robot.*, vol. 29, no. 21, pp. 1411–1425, Nov. 2015, doi: 10.1080/01691864.2015.1095652.

[28] M. Kaboli, P. Mittendorfer, V. Hugel, and G. Cheng, 'Humanoids learn object properties from robust tactile feature descriptors via multi-modal artificial skin', in *2014 IEEE-RAS International Conference on Humanoid Robots*, Madrid, Spain, Nov. 2014, pp. 187–192, doi: 10.1109/HUMANOIDS.2014.7041358.

[29] D. Xu, G. E. Loeb, and J. A. Fishel, 'Tactile identification of objects using Bayesian exploration', in *2013 IEEE International Conference on Robotics and Automation*, Karlsruhe, Germany, May 2013, pp. 3056–3061, doi: 10.1109/ICRA.2013.6631001.

[30] I. Bandyopadhyaya, D. Babu, A. Kumar, and J. Roychowdhury, 'Tactile sensing based softness classification using machine learning', in *2014*

IEEE International Advance Computing Conference (IACC), Gurgaon, India, Feb. 2014, pp. 1231–1236, doi: 10.1109/IAdCC.2014.6779503.

[31] N. Jamali and C. Sammut, 'Majority voting: Material classification by tactile sensing using surface texture', *IEEE Trans. Robot.*, vol. 27, no. 3, pp. 508–521, June 2011, doi: 10.1109/TRO.2011.2127110.

[32] Z. Yi, Y. Zhang, and J. Peters, 'Bioinspired tactile sensor for surface roughness discrimination', *Sens. Actuators Phys.*, vol. 255, pp. 46–53, Mar. 2017, doi: 10.1016/j.sna.2016.12.021.

[33] T. Bhattacharjee, J. M. Rehg, and C. C. Kemp, 'Haptic classification and recognition of objects using a tactile sensing forearm', in *2012 IEEE/RSJ International Conference on Intelligent Robots and Systems*, Vilamoura-Algarve, Portugal, Oct. 2012, pp. 4090–4097, doi: 10.1109/IROS.2012.6386142.

[34] H. Younes, A. Ibrahim, M. Rizk, and M. Valle, 'Data oriented approximate K-nearest neighbor classifier for touch modality recognition', presented at the 15th Conference on PhD Research in Microelectronics and Electronics, Lausanne, Switzerland, 2019.

[35] S. M. Afifi, H. Gholam Hosseini, and R. Sinha, 'Hardware implementations of SVM on FPGA: A state-of-the-art review of current practice', *International Journal of Innovative Science, Engineering & Technology*, vol. 2, no. 11, pp. 2348–7968, Nov. 2015.

[36] M. Kaboli, R. Walker, and G. Cheng, 'Re-using prior tactile experience by robotic hands to discriminate in-hand objects via texture properties', in *2016 IEEE International Conference on Robotics and Automation (ICRA)*, Stockholm, Sweden, May 2016, pp. 2242–2247, doi: 10.1109/ICRA.2016.7487372.

[37] J. A. K. Suykens and J. Vandewalle, 'Least squares support vector machine classifiers', *Neural Process. Lett.*, vol. 9, no. 3, pp. 293–300, 1999, doi: 10.1023/A:1018628609742.

[38] Z. Su et al., 'Force estimation and slip detection/classification for grip control using a biomimetic tactile sensor', in *2015 IEEE-RAS 15th International Conference on Humanoid Robots (Humanoids)*, Seoul, South Korea, Nov. 2015, pp. 297–303, doi: 10.1109/HUMANOIDS.2015.7363558.

[39] P. Gastaldo, L. Pinna, L. Seminara, M. Valle, and R. Zunino, 'A tensor-based pattern-recognition framework for the interpretation of touch modality in artificial skin systems', *IEEE Sens. J.*, vol. 14, no. 7, pp. 2216–2225, July 2014, doi: 10.1109/JSEN.2014.2320820.

[40] S. Decherchi, P. Gastaldo, R. S. Dahiya, M. Valle, and R. Zunino, 'Tactile-data classification of contact materials using computational intelligence', *IEEE Trans. Robot.*, vol. 27, no. 3, pp. 635–639, June 2011, doi: 10.1109/TRO.2011.2130030.

[41] S. Luo, W. Mou, M. Li, K. Althoefer, and H. Liu, 'Rotation and translation invariant object recognition with a tactile sensor', in *IEEE SENSORS 2014 Proceedings*, Valencia, Nov. 2014, pp. 1030–1033, doi: 10.1109/ICSENS.2014.6985179.

[42] 'Big-O Algorithm Complexity Cheat Sheet (Know Thy Complexities!) @ericdrowell'. https://www.bigocheatsheet.com/ (accessed Jan. 05, 2020).

[43] J. Schmidhuber, 'Deep learning in neural networks: An overview', *Neural Netw.*, vol. 61, pp. 85–117, Jan. 2015, doi: 10.1016/j.neunet.2014.09.003.

[44] A. E. Hassanien, A. Darwish, and C. L. Chowdhary, Eds., *Handbook of Research on Deep Learning Innovations and Trends*. IGI Global, Hershey, Pennsylvania 2019.

[45] D. J. Hemanth and V. V. Estrela, Eds., *Deep Learning for Image Processing Applications*. Amsterdam: IOS Press, 2017.

[46] J. M. Gandarias, A. J. García-Cerezo, and J. M. Gómez-de-Gabriel, 'CNN-based methods for object recognition with high-resolution tactile sensors', *IEEE Sens. J.*, vol. 19, no. 16, pp. 6872–6882, Aug. 2019, doi: 10.1109/JSEN.2019.2912968.

[47] O. Russakovsky et al., 'ImageNet Large Scale Visual Recognition Challenge', *ArXiv14090575 Cs*, Jan. 2015, [Online]. Available at http://arxiv.org/abs/1409.0575 [accessed Apr. 06, 2020].

[48] G. Rouhafzay and A.-M. Cretu, 'An application of deep learning to tactile data for object recognition under visual guidance', *Sensors*, vol. 19, no. 7, p. 1534, Jan. 2019, doi: 10.3390/s19071534.

[49] M. Alameh, A. Ibrahim, M. Valle, and G. Moser, 'DCNN for tactile sensory data classification based on transfer learning', in *2019 15th Conference on Ph.D Research in Microelectronics and Electronics (PRIME)*, Lausanne, Switzerland, July 2019, pp. 237–240, doi: 10.1109/PRIME.2019.8787748.

[50] H. Meng, K. Appiah, A. Hunter, and P. Dickinson, 'FPGA implementation of Naive Bayes classifier for visual object recognition', in *CVPR 2011 WORKSHOPS*, Colorado Springs, CO, USA, June 2011, pp. 123–128, doi: 10.1109/CVPRW.2011.5981831.

[51] G. Andrade et al., 'GPU-NB: A Fast CUDA-based implementation of Naive Bayes', in *2013 25th International Symposium on Computer Architecture and High Performance Computing*, Porto de Galinhas, Pernambuco, Brazil, Oct. 2013, pp. 168–175, doi: 10.1109/SBAC-PAD.2013.16.

[52] F. Saqib, A. Dutta, J. Plusquellic, P. Ortiz, and M. S. Pattichis, 'Pipelined decision tree classification accelerator implementation in FPGA (DT-CAIF)', *IEEE Trans. Comput.*, vol. 64, no. 1, pp. 280–285, Jan. 2015, doi: 10.1109/TC.2013.204.

[53] K. Jurczuk, M. Czajkowski, and M. Kretowski, 'Evolutionary induction of a decision tree for large-scale data: a GPU-based approach', *Soft Comput.*, vol. 21, no. 24, pp. 7363–7379, Dec. 2017, doi: 10.1007/s00500-016-2280-1.

[54] N. Attaran, A. Puranik, J. Brooks, and T. Mohsenin, 'Embedded low-power processor for personalized stress detection', *IEEE Trans. Circuits Syst. II Express Briefs*, vol. 65, no. 12, pp. 2032–2036, Dec. 2018, doi: 10.1109/TCSII.2018.2799821.

[55] M. Magno, A. Ibrahim, A. Pullini, M. Valle, and L. Benini, 'An energy efficient E-skin embedded system for real-time tactile data decoding', *J. Low Power Electron.*, vol. 14, no. 1, pp. 101–109, Mar. 2018, doi: 10.1166/jolpe.2018.1537.

[56] 'PULP platform'. https://www.pulp-platform.org/ (accessed Jan. 05, 2020).

[57] P. Mittendorfer and G. Cheng, 'Humanoid multimodal tactile-sensing modules', *IEEE Trans. Robot.*, vol. 27, no. 3, pp. 401–410, June 2011, doi: 10.1109/TRO.2011.2106330.

[58] P. Asman, T. Jiang, M. Ozturk, J. Reyna, and N. F. Ince, 'A low-cost microcontroller based stimulation system to study sensory processing', in *2019 9th International IEEE/EMBS Conference on Neural Engineering (NER)*, San Francisco, CA, USA, Mar. 2019, pp. 883–886, doi: 10.1109/NER.2019.8716944.

[59] 'TE0726 Zynqberry Demo1 – Public Docs – Trenz Electronic Wiki'. https://wiki.trenz-electronic.de/display/PD/TE0726+Zynqberry+Demo1 (accessed Jan. 05, 2020).

[60] M. Signoretto, L. De Lathauwer, and J. A. K. Suykens, 'A kernel-based framework to tensorial data analysis', *Neural Netw.*, vol. 24, no. 8, pp. 861–874, Oct. 2011, doi: 10.1016/j.neunet.2011.05.011.

[61] H. Younes, A. Ibrahim, M. Rizk, and M. Valle, 'Algorithmic level approximate computing for machine learning classifiers', presented at

the *26th IEEE International Conference on Electronics Circuits and Systems*, Genova, Italy, Nov. 2019.

[62] Y. Zhou, J. Lin, and Z. Wang, 'Energy efficient SVM classifier using approximate computing', in *2017 IEEE 12th International Conference on ASIC (ASICON)*, Guiyang, Oct. 2017, pp. 1045–1048, doi: 10.1109/ASICON.2017.8252658.

7

Approximate Arithmetic Circuits for Energy Efficient Data Processing in Electronic Skin

Mario Osta[1,*], Ali Ibrahim[2] and Maurizio Valle[1]

[1]DITEN-COSMIC Lab, University of Genova-Italy, Italy
[2]University of Genoa, Italy and Lebanese International University, Lebanon
E-mail: mario.osta@edu.unige.it; ali.ibrahim@edu.unige.it;
maurizio.Valle@unige.it
*Corresponding Author

The electronic (e-skin) system usually hosts many sensors coupled to an embedded electronic computing unit (ECU). The ECU extracts information by employing usually complex computing methods, e.g. machine learning (ML). Embedding ML algorithms poses challenges in terms of hardware resources and energy consumption because of (1) the high amount of data to be processed and (2) the computationally demanding computing methods. Leveraging on the trade-off between quality requirements versus computational complexity and time latency could reduce the system complexity without affecting the performance. In this chapter, we present the state-of-the-art approximate computing techniques (ACTs) employed at the circuit level for embedded ML algorithms and we prove the feasibility of the ACTs in the e-skin systems. Approximate arithmetic circuits, mainly multipliers and adders, are implemented in the embedded ML algorithms to enhance their efficiency. Approximate circuits can reduce the hardware complexity up to 16% at a cost of accuracy loss less than 5%.

7.1 Introduction

Energy-efficient circuits have become a substantial need for designing embedded computing systems for application domains such as wearable and

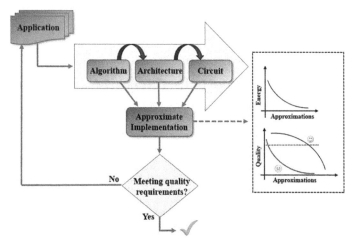

Figure 7.1   Design approach of approximate computing.

health-care devices. In particular, ECU is needed for the development of the
e-skin [1]. The e-skin system embeds autonomous and networked sensors;
each sensor hosts many input tactile sensors coupled to the ECU and power
supply [2]. The ECU locally extracts meaningful information by employing
usually complex computational methods, e.g., ML [3], and deals with large-
size datasets; however, this imposes challenges on the real-time operation and
adds a load on power consumption.

In recent years, the approximate computing paradigm has become a
significant major field of research since it may enhance the energy efficiency
and performance of digital systems. ACTs target error-resilient applications
to trade accuracy for power consumption, time latency, and hardware size
[4, 5]. ACTs can be applied at algorithmic, architectural, and circuit levels
as shown in Figure 7.1. Usually, approximate results are acceptable for many
applications, e.g., tactile data processing [3, 6] image processing [7], and data
mining [8]; thus, it is possible to take advantage of energy reduction with
minimal variation in performance [9].

In this perspective, the chapter presents the implementation of approximate arithmetic circuits in two blocks of the ECU: signal processing and data
decoding. The signal processing block is based on low-pass finite impulse
response (FIR) filters, and the data decoding employs ML algorithms, e.g.,
touch modality classification.

This chapter is organised as follows: Section 7.2 presents the state-of-
the-art of low power low-pass (FIR) filters. Section 7.3 presents the impact

of using approximate multipliers on FIR filter in the signal processing stage. Section 7.4 provides a summary of the state-of-the-art ACTs employed for embedded ML algorithms at the circuit level. Section 7.5 shows the implementation results obtained in terms of energy consumption and time latency after applying ACTs to the most expensive arithmetic circuits employed in data decoding. Section 7.6 summarizes the results.

7.2 Approximate Computing for Low-Pass Fir Filters

The e-skin system should acquire raw sensors' data, pre-process signals, and extract meaningful and relevant information. Usually, a digital low-pass FIR [10] filter reduces the noise contribution of the acquired signals. In this section, we introduce the state-of-the-art low power low-pass FIR filters for DSP applications and we show the implementation results of a case study dealing with six approximate FIR filters. The FIR filter circuit architecture [11] is shown in Figure 7.2. The operation of $N - tap$ FIR filter with input $x(n)$, output $y(n)$, and filter coefficients $H(m)$ is shown as

$$y(n) = \sum_{m=0}^{N-1} H(m) \times x(n - m) \tag{7.1}$$

Each input in the FIR filter is multiplied by the filter coefficients and then the results are added together to generate the output; this results in a high number of operations, i.e., mainly multiplications and additions.

Most of the proposed works in the state-of-the-art approximate FIR filters concentrate on minimizing or eliminating the multipliers in the FIR filters. Multipliers are substituted by multiple constant multipliers (MCM) having only shift-and-add operations [12]. The cost can be reduced when combining the MCMs with a canonical signed digit (CSD) [12], or with common subexpression elimination (CSE) [13]. On the other hand, different techniques are proposed to reduce the critical path for high throughput. In Ref. [14], weights

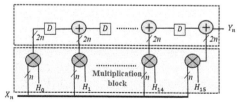

Figure 7.2 Structure of low pass finite impulse response filter.

Table 7.1   Comparison of various FIR designs [18]

FIR Filters	Number of Taps	Bit Widths	Adders
Approximate FIR_108 [20]	108	9	4bit-truncated n
Approximate FIR_279 [21]	279	12	6 bit-truncated
Approximate FIR_119 [19]	119	16	11 bit-truncated
Approximate FIR_61 [19]	61	16	12 bit-truncated
Approximate FIR_28 [19]	28	16	13 bit-truncated
Approximate FIR_13 [19]	13	16	14 bit-truncated

are employed as addresses to access the LUT which implements the addition operation. To improve efficiency, a more efficient DA is suggested in Ref. [15] for the block least mean square algorithm (BLMS). Power/area/latency is improved when updating the LUT process. However, the size of the LUT is large and increases area and power consumption. DA-based FIR filters perform better for low-order filters. For high-order filters, DA architectures are more efficient when employing decomposition techniques. In Ref. [16], an implementation which consists of shared LUTs is proposed to implement the DA for a reconfigurable FIR filter. A reduction in energy consumption (15%) is observed when compared to the systolic decomposition of a DA-based design.

However, Refs. [17, 18] show that ACTs outperform the low-power techniques presented previously for FIR. Since multiplication and addition account for more than 50% of the area and power consumption [17], implementing approximate multipliers and adders is considered a promising approach. Further, Ref. [18] shows that employing ACTs on FIR designs reduces the area by 35% and the power consumption by around 27%.

In the following, we present the result of a case study [18] concerning the implementation of six approximate CSD-FIR filters [19–21] using truncated adders. Table 7.1 shows the specifications of each FIR filter. Figure 7.3 shows the area/energy reduction of each approximated FIR filter. The reason behind implementing truncated adders in the FIR filter is since the least significant bits (LSBs) in the adder could be omitted, the LSBs have little impact on the output [18]. According to Ref. [18], the number of truncated bits (k) is computed in such a way that the average error does not exceed 1 *ulp* (*unit of least precision*). The *ulp* unit is used for error analysis for FIR filter [22]. Figure 7.3 shows that the approximate FIR_13 and approximate FIR_119 achieve, respectively, an area and power reduction by up to 35%

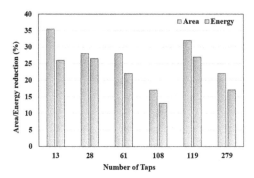

Figure 7.3 Percentage of area and energy reduction of FIR filters for different numbers of taps [18].

and 27%. Among the four FIR filters having the same bit width (16 bits), the approximate FIR_19 has the highest power consumption reduction (27%) since the number of truncated bits decreases with the increasing of filter order.

7.3 Approximate Filters for E-skin

This section presents the implementation results of approximate multipliers from the state-of-the-art FIR filter in the e-skin. A fully parallel 16-tap low-pass (FIR) filter based on transposed form architecture (16 bits dynamic range) [23], is employed in the e-skin system. Approximate multipliers are adopted in the FIR filter to reduce its complexity. The FIR filters are implemented on Virtex-7 xc7vx485tffg1157-1.

In the following, we present the impact of using ACTs on the FIR filter and how such techniques affect the results in terms of signal-to-noise ratio (SNR) and power consumption.

The SNR is calculated with reference to the exact filter which employs the exact Baugh-Wooley multiplier [24]. Table 7.2 shows the results for six different approximate multipliers. Among the five approximate multipliers [25–28] achieve the best performance in terms of SNR. However, Ref. [25] shows respectively an improvement of power consumption and area by 45% and 13% with respect to Ref. [24]. The power and area of Ref. [26] are reduced by 35% and 15%, respectively. However, BW-AXA reached values (i.e. 3 dB) far from being accepted for the application. Applying ACTs on the FIR filter could achieve an improvement of power consumption by an average of more than 40% at the cost of less than 0.3 dB degradation in the SNR.

Table 7.2   Signal-to-noise ratio, power, and area reduction of FIR based multipliers

FIR-based Multipliers	SNR (dB)	Power Reduction (%)	Area Reduction (%)
Exact BW [23]	23.3	–	–
Kulk [24]	23.2	45	13
ROBA [25]	23	65	15
Evo25 [26]	19.2	68.5	10
META [27]	16.8	48	15
Shaf [26]	15.7	49.5	20

7.4 Approximate Computing for Embedded Machine Learning

This section assesses circuit level ACTs employed in the literature for the embedded implementation of ML algorithms. Figure 7.4 introduces graphically main techniques, i.e., approximate arithmetic circuits, approximate memory, and quantization [29].

7.4.1 Approximate Arithmetic Circuits

ML algorithms require a high number of multiplications, which increases the complexity of the hardware implementation. Introducing approximate multipliers in the embedded ML implementation could reduce meaningfully the circuit complexity. Table 7.3 shows the performance results of the ACTs techniques for different ML applications. Several approximate multipliers for the ML algorithms are proposed in the literature [8, 30–32]. However, Ref. [30] evaluated the use of an alphabet set multiplier (ASM) in a deep neural network, i.e., the conventional multiplication is substituted by simplified shift and add operations [30]. The power consumption is reduced from 18% to 27% at the cost of an accuracy loss of less than 0.4%. In Ref. [31],

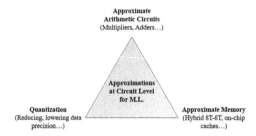

Figure 7.4   Approximate computing techniques for embedded ML at circuit level.

Table 7.3 Comparison of approximate computing techniques for embedded machine learning algorithms at circuit level

Technique	Design Approach	ML and DL Algorithms	Performance	Accuracy Degradation
Approximate Multipliers	Add and shift operations [30]	DNN	18% till 27% energy savings	<0.5%
	Inexact logic minimization approach [31]	NN	44% till 62% power savings	MSE = 2
	Approximate BW [63]	SVM based tensorial kernel approach	16% energy savings	<5%
Approximate Adders	IMPACT adders [33]	CNN	30% power savings	14%
	LOA adders [32]	NN	50% area delay product reduction	
Approximate Memory	Hybrid 8T-6T SRAM cell [30]	DCNN	0.5 V reduction of the operating voltage	<0.6%
	Approximate on-chip caches [36]	CNN	1.3× speedup improvement	1%
	Approximate memory compression [41]	8 K-means, Fuzzy K-means	1.2× energy savings and 11.5% reduction in time	1.5%
	Linearly compressed pages [40]	NN	9.5% energy reduction	5%
Quantization	Reduced precision data [42]	CNN	2× energy savings	1%
	Quantizer design [44, 47]	NN	20% model reduction	No accuracy loss
	Lowering the precision [45]	DCNN	5× reduction of power and area	1%

Deep Neural Network (DNN), Neural Network (NN), Support Vector Machine (SVM), Convolutional Neural Network (CNN), Deep Convolutional Neural Network (DCNN).

the energy efficiency is improved by 43.9% to 62.5% after implementing the approximate multiplier using the inexact logic minimization approach in a neural network.

The addition is also a fundamental operation and has a significant influence on the performance of the computational system. In Ref. [33], the authors have selected five configurations for one approximate adder called IMPACT adder. The power consumption is improved by 30% when the approximate adder is introduced into a deep convolutional neural network (CNN) architecture based on LeNet-5 [34]. In Ref. [32], authors proposed the use of lower-part-OR adder (LOA) and the broken-array multiplier (BAM) in a three-layer neural network for face recognition and in the hardware defuzzification block of a fuzzy processor. The power delay product decreased by around 50%.

7.4.2 Approximate Memory

Memories in the embedded electronic systems have a significant impact on their energy consumption. Optimizing the memory cells improves meaningful energy efficiency [35]. We will introduce some examples. The authors in Ref. [30] have applied the approximations to weighted synapses that interconnect different layers and proposed a hybrid 8T-6T SRAM cell. The MSBs are stored in the 8T bit-cells while the LSBs are stored in the 6T bit–cells. After implementing the hybrid 8T-6T SRAM cell in a deep, fully connected network, the operating voltage is reduced from 0.85 V to 0.8 V at the cost of an accuracy degradation of less than 0.5% for 12-bit synaptic weight.

Some works have focused on approximations within on-chip caches such as in Ref. [36], where a novel asymmetric compute-memory extension (ACME) is introduced. The ACME consists in modifying the format of data in the memory by removing the marginal bits. This reduces the cost of storing and moving bits throughout the memory hierarchy. The speed is improved by $1.3\times$ at the cost of accuracy loss of less than 1%. In Refs. [37, 38], the energy efficiency is improved by proposing a quality-configurable memory array and a STAxCache called (spintronic approximate cache). Quality configurable memory aims to store data at different levels based on the requirement of each application. STAxCache is based on approximate L2 cache architecture, which aims to retain the full flexibility of a conventional cache. In Ref. [37] the energy efficiency is improved by 19.5% at the cost of less than 0.5% of accuracy loss, whereas in Ref. [38] the energy is improved by $1.44\times$ in the L2 cache with same accuracy loss.

Other efforts have been focusing on memory compression, such as in Ref. [39], where the authors introduced a memory compression scheme to a DNN architecture to reduce its energy consumption. In Ref. [40] the authors have proposed a technique called linearly compressed pages (LCP), which compresses all of the cache lines within a given page to the same size. The energy efficiency of the main memory subsystem is decreased by 9.5%. Based on the memory compression approach mentioned earlier, the authors in Ref. [41] have enhanced the memory controller to be cautious of the memory regions containing approximation resilient data. Data written-to/read-from these regions is compressed and decompressed. This technique is applied to 8 ML benchmarks and improvement of 1.28× in DRAM energy and 11.5% in time execution is achieved, respectively, for less than 1.5% accuracy loss.

7.4.3 Quantization

Quantization is used as a hardware approximation technique: the floating-point operations are substituted by fixed point ones. Fixed-point implementation is explored in different works. The authors in Refs. [42–46] have used different data formats representation such as floating point, fixed point, and dynamic fixed point to train the neural network. They showed that low or limited precision data can be used for training and inference.

Low precision data techniques are applied in different architectures such as the DNN and CNN. A reduction of data footprint by an average of 74% up to 92% at the cost of an accuracy loss of 1%, in the CNN, is shown in Ref. [43]. In deep-learning accelerators (e.g., dLAC), the low precision (2-bit) weight approach adopted in Ref. [45] improved the power and area of the accelerator by 5×.

The quantizer design proposed in Ref. [44] for the CNN reduced the model size by 20% without any loss in accuracy on the CIFAR-10 benchmark. However, Gysel et al. [47] proposed a framework called 'Ristretto', which reduces the size of the models by employing fixed-point representation instead of floating point and by re-training of the CNNs based on the Caffe framework [48]. They showed that the storage requirements and the multiplier energy could be reduced, respectively, by up to 36% and 50%. The LookNN approach [49], which aims to substitute the floating-point multiplications with lookup tables, succeeded in improving the energy and the speed by 3× and 2.6×, respectively, when compared to the traditional GPU architecture implementation.

7.5 Approximate Embedded Machine Learning for E-skin

ML algorithms provide powerful solutions for non-linear and complex problems through a 'learning by examples' approach. Such methods are employed to design predictive systems that can make decisions on unseen input samples [50, 51]. In the case of data decoding, ML methods have been investigated

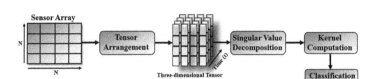

Figure 7.5 Functional steps of tensorial kernel approach.

to extract structured information from raw data, e.g., texture/touch modality classification. For the touch modality classification task, a tensorial kernel approach to preserve the inherent structure of the original data in the e-skin system has been proposed [52].

In this section, we will briefly describe the tensorial kernel approach. Then, we will discuss the improvements obtained in terms of energy consumption and time latency after applying ACTs to the most expensive arithmetic circuits.

7.5.1 Tensorial Kernel Approach

Usually, ML approaches are not employed to handle data in tensor form. Figure 7.5 illustrates the functional blocks of the approach proposed in Ref. [52]. The input tactile data generated from the sensor array are arranged in a three-dimensional tensor through the tensor arrangement block. The singular value decomposition (SVD) computes the eigenvectors of the three matrices unfolded from the tensor. Then, the classification block classifies touch modalities after computing the kernel function.

The SVD is a complex circuit [53]. To assess the computational cost of the tensorial kernel approach, a case study in Ref. [54] has been taken into consideration. The number of operations needed for real-time single classification has been evaluated in Ref. [54]. Based on the reported results, classifying three touch modalities poses a tough challenge for the e-skin system.

7.5.2 Coordinate Rotational Digital Computer Circuits

CORDIC circuit is based on the concept of rotational geometry as described by Volder [55, 56]. The complexity lies in the computation of the angle of rotation (i.e., cosine and sine functions) and in the management of the rotations. CORDIC is a block of the SVD. The following section introduces the CORDIC algorithm and describes the steps followed to apply the approximate computing techniques.

7.5.2.1 CORDIC algorithm

CORDIC algorithm consists of a sequence of shift-add operations to compute several functions. CORDIC is usually employed in two modes: vectoring and rotating. In the vectoring mode, the input vector is rotated by any angle which aligns the resultant vector with the horizontal axis. In the rotation mode, the desired rotation angle initializes the angle accumulator. The rotation decision criteria (d_i) at each iteration diminishes the magnitude of the residual angle in the angle accumulator. The decision at each iteration is based on the sign of the residual angle. The iteration equations are given by

$$X_{i+1} = X_i - Y_i \times d_i \times 2^{-i}$$
$$Y_{i+1} = Y_i + X_i \times d_i \times 2^{-i} \qquad (7.2)$$
$$Z_{i+1} = Z_i - d_i \times tan^{-1}(2^{-i})$$

where

$$\begin{cases} d_i = +1 & \text{if } Y_i < 0, -1 \text{ otherwise for vectoring mode} \\ & \text{and} \\ d_i = -1 & \text{if } Z_i < 0, +1 \text{ otherwise for rotation mode} \end{cases} \qquad (7.3)$$

they provide the following results:

* Rotation mode:

$$X_n = A_n[X_0 cos Z_0 - Y_0 sin Z_0]$$
$$Y_n = A_n[Y_0 cos Z_0 - X_0 sin Z_0] \qquad (7.4)$$
$$Z_n = 0$$

* Vectoring mode

$$X_n = A_n\sqrt{X_0^2 + Y_0^2}$$
$$Y_n = 0 \qquad (7.5)$$
$$Z_n = Z_0 + tan^{-1}\left(\frac{Y_0}{X_0}\right)$$

7.5.2.2 Approximate CORDIC implementation

CORDIC uses a single shift-add operation for each component: x, y, and z, as shown in Figure 7.6. A MUX (2:1 multiplexer), a shift register, and an adder/subtractor are required for each unit as shown in Figure 7.6. Before the beginning of computation, three input X_{in}, Y_{in}, and Z_{in} values are stored in

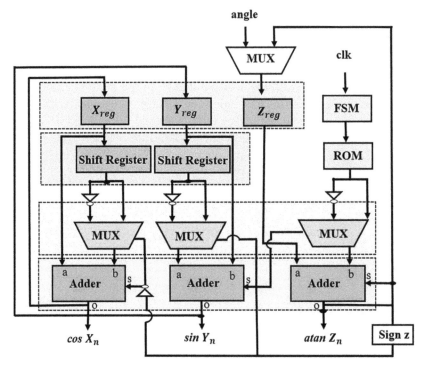

Figure 7.6   CORDIC architecture.

X_{reg}, Y_{reg}, and Z_{reg}. The micro-rotation angles arctan 2^{-i} are stored in the ROM, where the input of the ROM i changes from 0 to 29 [57]. The FSM is responsible for (1) enabling the multiplexer signals and (2) controlling the ROM addresses. Addition and subtraction are performed simultaneously. The current operation (addition or subtraction) is based on the sign of Z_{reg}. The $cosX_n$ and $sinY_n$ resulting from the CORDIC [58] computation are computed after setting $y_0 = 0$ and $x_0 = \frac{1}{A_n}$ where $A_n = 0.6073$ and n is the index.

As shown in Figure 7.6, the exact CORDIC architecture is composed of three ripple carry adders (exact adders) [59]; in this case study, approximate adders are implemented.

The steps to implement the approximate CORDIC circuit in the case study are described hereunder:

(1) Efficient approximate adder circuits from the literature have been selected. The approximate adders belong to two classes: speculative adders and approximate full adders.

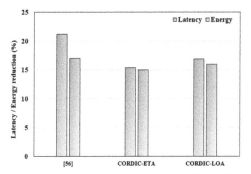

Figure 7.7   Power and latency reductions of CORDIC circuit.

- Speculative adders : (approximate XNOR-based adder (AXA) [60], input pre-processing [61]).
- Approximate full adders: (approximate NAND-carry out bit [62], approximate AND-carry out bit [62], (LOA) [32], and error tolerant adder (ETA) [63]).

(2) Implementation of the approximate adders on Virtex-7 xc7vx485tffg 1157-1
(3) Evaluation of the performance
(4) Choosing the approximate adders in terms of accuracy and energy and implementing them into the CORDIC block

The ETA and LOA have been selected. Figure 7.7 shows that the energy consumption of CORDIC-ETA and CORDIC-LOA is reduced, respectively, by 15% and 15.4% with respect to the exact CORDIC.

7.5.3 Singular Value Decomposition

The singular value decomposition (SVD) has been introduced by Jordan and Betrami [64] for real square matrices. The SVD is a matrix factorization method employed to analyse the structure and properties of a matrix. The SVD is used to obtain the least square solutions of a system of linear equations [65]. In the following subsection, we will describe the SVD algorithm and the hardware implementation as shown in Figure 7.8.

7.5.3.1 SVD algorithm
The SVD is used to factorize a matrix M of size $(M \times N)$ into a product as shown hereunder:

$$M = USV^T \qquad (7.6)$$

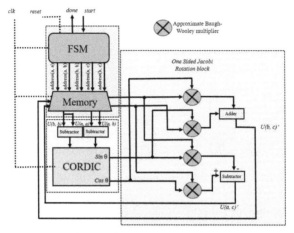

Figure 7.8   SVD block diagram.

where $U, S,$ and V are, respectively, an orthogonal matrix $(m \times m)$, a diagonal matrix $(m \times n)$, and a unitary matrix $(n \times n)$. The elements in the columns of U and V are, respectively, the left- and right singular vectors of the matrix M. The singular values are represented by the diagonal elements $(\sigma_0, \ldots, \sigma_{n-1})$. In this study, the SVD is implemented based on the one-sided Jacobi algorithm [66], which consists of applying a sequence of rotations to the matrix $U = M^T M$. The algorithm requires eight iterations to reach the convergence. The sequence $U_1, U_2, U_3 \ldots$ is generated by computing U_{a+1} as follows:

$$U_{a+1} = J(a, b, \theta)^T U_a \, J(a, b, \theta) \tag{7.7}$$

where $J(a, b, \theta)$ is the Jacobi rotation composed of an identity square matrix equal I with four elements on the intersection of rows a and columns b. A Jacobi rotation is computed for each sub-matrix of size (2×2) to annihilate the off-diagonal elements (w) of the matrix U as follows:

$$\begin{bmatrix} \hat{x} & 0 \\ 0 & \hat{y} \end{bmatrix} = \begin{bmatrix} \cos\theta & -\sin\theta \\ \sin\theta & \cos\theta \end{bmatrix}^T \begin{bmatrix} x & w \\ w & y \end{bmatrix} \begin{bmatrix} \cos\theta & -\sin\theta \\ \sin\theta & \cos\theta \end{bmatrix} \tag{7.8}$$

7.5.3.2 Approximate SVD

The block diagram for the implementation of the SVD algorithm based on the one- sided Jacobi is shown in Figure 7.8. SVD requires four multipliers for the one-sided Jacobi rotation block. To reduce the complexity, four approximate

Table 7.4 Percentage relative error of approximate SVD

Eigenvalues	Relative Error (%)		
	SVD-approx8	SVD-approx12	SVD-approx16
S1	0	0	0.018
S2	0	0	0.097
S3	0	0.054	0.27
S4	0	0.31	0.62
S5	0	0	1.25

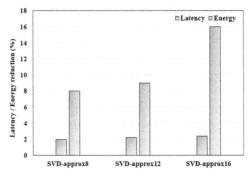

Figure 7.9 Latency and energy reductions of SVD circuit.

Baugh-Wooley multipliers [63] are implemented, while adders and subtractors are exact. The scalability of the approximate multiplier is assessed by the SVD by approximating eight LSBs, and then increasing the number of approximated bits reaching 16 approximated LSBs. The error-resilience [67] of the approximate SVD is evaluated based on the relative error (RE) defined as follows:

$$RE(\%) = \frac{|\text{Approximate result} - \text{Exact result}|}{\text{Exact result}} \times 100 \qquad (7.9)$$

We evaluated the performance of the approximate SVD in terms of latency and energy reduction in a benchmark with matrices of size 5×5. The SVD is implemented in Vivado using VHDL Language on Virtex-7 xc7vx485tffg1157-1. Table 7.4 shows the RE of the eigenvalues generated from the approximate SVD with respect to the exact SVD after applying the approximations for different numbers of bits.

The eigenvalues of the SVD-approx12 and SVD-approx16 reach, respectively, an average RE of around 0.07% and 0.45%, i.e., the RE increases when increasing the number of approximated LSBs of the approximate BW.

Table 7.5 Error rate of the SVM based tensorial kernel approach for different values of approximate LSBs in the SVD

	Error Rate (%)	
SVD	Sliding	Rolling
Exact-SVD	12.5	20
SVD-approx8	12.5	20
SVD-approx12	12.5	20
SVD-approx16	13.75	20

The latency and energy consumption are reduced, respectively, by up to 2.4% and 16% after approximating 16 LSBs as shown in Figure 7.9.

7.6 Discussion and Conclusion

In this case study, the approximate SVD circuit implementation as described earlier has been modelled in C language and implemented in the SVM tensorial kernel algorithm. The algorithm has been simulated in a case study as follows.

The dataset is the one of Ref. [6]: 70 participants were asked to apply two touch modalities (sliding the finger vs rolling a washer) on a 16-sensors array. Each touch modality was repeated twice for two directions (horizontally and vertically). The total number of collected touch modalities is 560 (70 participants, 2 modalities, 2 directions, 2 trials).

The ER is computed as follows:

$$ER(\%) = \frac{\text{number of incorrect touch modalities}}{\text{total number of touch modalities}} \times 100 \qquad (7.10)$$

The ER of sliding and rolling are, respectively, equal to 12.5% and 20% when simulating the exact SVD in the SVM as shown in Table 7.5. These results are used as a baseline to assess the impact of the approximate SVD on the SVM.

The ER in the case of SVD-approx8 and SVD-approx12 with respect to the exact SVD is fairly constant; the ER in the case of SVD-approx16 is increased by 1.25% for sliding.

The chapter assessed the application of ACTs at the circuit level for embedded ML in the e-skin systems. Results demonstrated that using approximate multipliers in the FIR filter led to a power reduction of around 80% with an SNR degradation of 1.39 dB. We applied ACTs to the CORDIC and

SVD, which are the main bottlenecks of the embedded ML implementation. We evaluated the impact of the approximate SVD in the SVM. Results demonstrated the feasibility of the ACTs for the tensorial SVM algorithm for classifying touch modalities (sliding vs rolling) by increasing the ER only by 1.25% while reducing the power consumption of the SVD by 16%. The implementation of different ACTs such as approximate memory storage and voltage-scaled memory (circuit level) [30, 36] and quantization techniques (algorithmic level) [44, 45] should be investigated.

References

[1] M. Franceschi, L. Seminara, S. Dosen, M. Strbac, M. Valle, and D. Farina, "A system for electrotactile feedback using electronic skin and flexible matrix electrodes: Experimental evaluation," *IEEE Transactions on Haptics*, vol. 10, no. 2, pp. 162–172, 1 April–June 2017. doi: 10.1109/TOH.2016.2618377

[2] M. Magno, A. Ibrahim, A. Pullini, M. Valle, and L. Benini, "Energy efficient system for tactile data decoding using an ultra-low power parallel platform," 2017 New Generation of CAS (NGCAS), Genova, pp. 17–20, 2017. doi: 10.1109/NGCAS.2017.56

[3] A. Ibrahim, P. Gastaldo, H. Chible, and M. Valle, "Real-time digital signal processing based on FPGAs for electronic skin implementation," *Sensors*, vol. 17.3, p. 558, 2017.

[4] S. Mittal, "A survey of techniques for approximate computing," *ACM Computing Surveys (CSUR)*, vol. 48.4, pp. 1–33, 2016.

[5] J. Han and M. Orshansky, "Approximate computing: An emerging paradigm for energy-efficient design," in 2013 18th IEEE European Test Symposium (ETS), Avignon, pp. 1–6, 2013. doi: 10.1109/ETS.2013.6569370

[6] P. Gastaldo, L. Pinna, L. Seminara, M. Valle, and R. Zunino, "A tensor-based approach to touch modality classification by using machine learning," *Robotics and Autonomous Systems*, vol. 63, 268–278, 2015.

[7] R. Seva, P. Metku, K. K. Kim, Y. Kim, and M. Choi, "Approximate stochastic computing (ASC) for image processing applications," in 2016 International SoC Design Conference (ISOCC), Jeju, pp. 31–32, 2016. doi: 10.1109/ISOCC.2016.7799758

[8] Q. Zhang, T. Wang, Y. Tian, F. Yuan, and Q. Xu, "ApproxANN: An approximate computing framework for artificial neural network," 2015

Design, Automation & Test in Europe Conference & Exhibition (DATE), Grenoble, pp. 701–706, 2015.

 [9] D. Mohapatra, G. Karakonstantis, and K. Roy, "Significance driven computation: A voltage-scalable, variation-aware, quality-tuning motion estimator," in *Proceedings of the International Symposium on Low Power Electronics and Design*, pp. 195–200, 2009.

[10] L. Seminara et al., "Towards integrating intelligence in electronic skin," *Mechatronics*, vol. 34, pp. 84–94, 2016.

[11] B. A. Shenoi, *Introduction to Digital Signal Processing and Filter Design*. John Wiley & Sons, Hoboken, New Jersey, 2005.

[12] N. Goel and A. Nandi, "Design of FIR filter using FCSD representation," in 2015 IEEE International Conference on Computational Intelligence & Communication Technology, Ghaziabad, pp. 617–620, 2015.

[13] W. B. Ye, X. Lou, and Y. J. Yu, "Design of low-power multiplierless linear-phase FIR filters," *IEEE Access*, vol. 5, pp. 23466–23472, 2017.

[14] R. Guo and L. S. DeBrunner, "Two high-performance adaptive filter implementation schemes using distributed arithmetic," *IEEE Transactions on Circuits and Systems II: Express Briefs,* vol. 58, no. 9, pp. 600–604, Sept. 2011. doi: 10.1109/TCSII.2011.2161168

[15] B. K. Mohanty and P. K. Meher, "A high-performance energy-efficient architecture for FIR adaptive filter based on new distributed arithmetic formulation of block LMS algorithm," *IEEE Transactions on Signal Processing,* vol. 61, no. 4, pp. 921–932, Feb. 15, 2013. doi: 10.1109/TSP.2012.2226453

[16] S. Y. Park and P. K. Meher, "Efficient FPGA and ASIC realizations of a DA-based reconfigurable FIR digital filter," *IEEE Transactions on Circuits and Systems II: Express Briefs*, vol. 61, no. 7, pp. 511–515, July 2014. doi: 10.1109/TCSII.2014.2324418

[17] X. Lou, Y. J. Yu, and P. K. Meher, "Analysis and optimization of product-accumulation section for efficient implementation of FIR filters," *IEEE Transactions on Circuits and Systems I: Regular Papers,* vol. 63, no. 10, pp. 1701–1713, Oct. 2016. doi: 10.1109/TCSI.2016.2587105

[18] J. Ye, N. Togawa, M. Yanagisawa, and Y. Shi, "Static error analysis and optimization of faithfully truncated adders for area-power efficient FIR designs," in 2019 IEEE International Symposium on Circuits and Systems (ISCAS), Sapporo, Japan, pp. 1–4, 2019. doi: 10.1109/ISCAS.2019.8702386

[19] D. L. Maskell, "Design of efficient multiplierless FIR filters," *IET Circuits, Devices & Systems,* vol. 1, no. 2, pp. 175–180, Apr. 2007.

[20] S. Mirzaei, R. Kastner, and A. Hosangadi, "Layout aware optimization of high speed fixed coefficient FIR filters for FPGAs," *International Journal of Reconfigurable Computing,* vol. 2010, no. 1, p. 17, Jan. 2010.

[21] R. Mahesh and A. P. Vinod, "A new common subexpression elimination algorithm for realizing low-complexity higher order digital filters," *IEEE Transactions on Computer-Aided Design of Integrated Circuits and Systems,* vol. 27, no. 2, pp. 217–229, Feb. 2008. doi: 10.1109/TCAD.2007.907064

[22] S. Hsiao, J. Zhang Jian, and M. Chen, "Low-cost FIR filter designs based on faithfully rounded truncated multiple constant multiplication/accumulation," *IEEE Transactions on Circuits and Systems II: Express Briefs,* vol. 60, no. 5, pp. 287–291, May 2013. doi: 10.1109/TCSII.2013.2251958

[23] V. S. Rosa, F. F. Daitx, E. Costa, and S. Bampi, "Design flow for the generation of optimized FIR filters," in 2009 16th IEEE International Conference on Electronics, Circuits and Systems – (ICECS 2009), Yasmine Hammamet, pp. 1000–1003, 2009. doi: 10.1109/ICECS.2009.5410839

[24] V. S. Muley, A. Tom, and Vigneswaran T., "Design of Baugh Wooley and Wallace tree multiplier using two phase clocked adiabatic static CMOS logic," in 2015 International Conference on Industrial Instrumentation and Control (ICIC), Pune, pp. 1178–1183, 2015. doi: 10.1109/IIC.2015.7150926

[25] P. Kulkarni, P. Gupta, and M. D. Ercegovac, "Trading accuracy for power in a multiplier architecture," *Journal of Low Power Electronics,* vol. 7.4, 490–501, 2011.

[26] R. Zendegani, M. Kamal, M. Bahadori, A. Afzali-Kusha, and M. Pedram, "RoBA multiplier: A rounding-based approximate multiplier for high-speed yet energy-efficient digital signal processing," *IEEE Transactions on Very Large Scale Integration (VLSI) Systems,* vol. 25, no. 2, pp. 393–401, Feb. 2017. doi: 10.1109/TVLSI.2016.2587696

[27] V. Mrazek, R. Hrbacek, Z. Vasicek, and L. Sekanina, "EvoApproxSb: Library of approximate adders and multipliers for circuit design and benchmarking of approximation methods," in *Proceedings of the 2017 Design, Automation and Test in Europe, DATE 2017, IEEE,,* pp. 258–261, 2017.

[28] M. Osta, A. Ibrahim, L. Seminara, H. Chible, and M. Valle, "Low power approximate multipliers for energy efficient data processing," *Journal of Low Power Electronics*, vol. 14.1, pp. 110–117, 2018.

[29] M. Osta, M. Alameh, H. Younes, A. Ibrahim, and M. Valle, "Energy efficient implementation of machine learning algorithms on hardware platforms," in 2019 26th IEEE International Conference on Electronics, Circuits and Systems (ICECS), Genoa, Italy, pp. 21–24, 2019. doi: 10.1109/ICECS46596.2019.8965157

[30] S. S. Sarwar et al., "Energy efficient neural computing: A study of cross-layer approximations," *IEEE Journal on Emerging and Selected Topics in Circuits and Systems*, vol. 8, no. 4, pp. 796–809, Dec. 2018. doi: 10.1109/JETCAS.2018.2835809

[31] Z. Du, A. Lingamneni, Y. Chen, K. V. Palem, O. Temam, and C. Wu, "Leveraging the error resilience of neural networks for designing highly energy efficient accelerators," *IEEE Transactions on Computer-Aided Design of Integrated Circuits and Systems*, vol. 34, no. 8, pp. 1223–1235, Aug. 2015. doi: 10.1109/TCAD.2015.2419628

[32] H. R. Mahdiani, A. Ahmadi, S. M. Fakhraie, and C. Lucas, "Bio-inspired imprecise computational blocks for efficient VLSI implementation of soft-computing applications," *IEEE Transactions on Circuits and Systems I: Regular Papers,* vol. 57, no. 4, pp. 850–862, April 2010. doi: 10.1109/TCSI.2009.2027626

[33] M. Shafique et al., "Adaptive and energy-efficient architectures for machine learning: Challenges, opportunities, and research roadmap," in 2017 IEEE Computer Society Annual Symposium on VLSI (ISVLSI), Bochum, pp. 627–632, 2017. doi: 10.1109/ISVLSI.2017.124

[34] V. Sze, Y. Chen, T. Yang, and J. S. Emer, "Efficient processing of deep neural networks: A tutorial and survey," in *Proceedings of the IEEE*, vol. 105, no. 12, pp. 2295–2329, Dec. 2017. doi: 10.1109/JPROC.2017.2761740

[35] M. Shafique, F. Sampaio, B. Zatt, S. Bampi, and J. Henkel, "Resilience-driven STT-RAM cache architecture for approximate computing," in *Workshop on Approximate Computing (AC)*, Paderborn, Germany, October 2015.

[36] A. Jain et al., "Concise loads and stores: The case for an asymmetric compute-memory architecture for approximation," in 2016 49th Annual IEEE/ACM International Symposium on Microarchitecture (MICRO), Taipei, pp. 1–13, 2016. doi: 10.1109/MICRO.2016.7783744

[37] A. Ranjan, S. Venkataramani, X. Fong, K. Roy, and A. Raghunathan, "Approximate storage for energy efficient spintronic memories," in 2015 52nd ACM/EDAC/IEEE Design Automation Conference (DAC), San Francisco, CA, pp. 1–6, 2015. doi: 10.1145/2744769.2744799

[38] A. Ranjan, S. Venkataramani, Z. Pajouhi, R. Venkatesan, K. Roy, and A. Raghunathan, "STAxCache: An approximate, energy efficient STT-MRAM cache," in Design, Automation & Test in Europe Conference & Exhibition (DATE), 2017, Lausanne, pp. 356–361, 2017. doi: 10.23919/DATE.2017.7927016

[39] P. Judd, J. Albericio, T. Hetherington, T. M. Aamodt, N. E. Jerger, and A. Moshovos, "Proteus: Exploiting numerical precision variability in deep neural networks," in *Proceedings of the 2016 International Conference on Supercomputing*, pp. 1–12, vol. 01–03–June, Istanbul Turkey, 2016.

[40] G. Pekhimnko et al., "Linearly compressed pages: A low-complexity, low-latency main memory compression framework," in 2013 46th Annual IEEE/ACM International Symposium on Microarchitecture (MICRO), Davis, CA, Dec, pp. 172–184, 2013.

[41] A. Ranjan, A. Raha, V. Raghunathan, and A. Raghunathan, "Approximate memory compression for energy-efficiency," in 2017 IEEE/ACM International Symposium on Low Power Electronics and Design (ISLPED), Taipei, pp. 1–6, 2017. doi: 10.1109/ISLPED.2017.8009173

[42] S. Gupta, A. Agrawal, K. Gopalakrishnan, and P. Narayanan, "Deep learning with limited numerical precision," in 32nd International Conference on Machine Learning, June 1, pp. 1737–1746, ICML 2015.

[43] P. Judd et al., "Reduced-precision strategies for bounded memory in deep neural nets," arXiv preprint arXiv:1511.05236, 2015.

[44] D. D. Lin, S. S. Talathi, and V. S. Annapureddy, "Fixed point quantization of deep convolutional networks," in 33rd International Conference on Machine Learning, ICML 2016, pp. 2849–2858, 2016.

[45] G. Venkatesh, E. Nurvitadhi, and D. Marr, "Accelerating deep convolutional networks using low-precision and sparsity," in 2017 IEEE International Conference on Acoustics, Speech and Signal Processing (ICASSP), New Orleans, LA, pp. 2861–2865, 2017. doi: 10.1109/ICASSP.2017.7952679

[46] M. Courbariaux, J. P. David, and Y. Bengio, "Training deep neural networks with low precision multiplications," in 3rd International Conference on Learning Representations, ICLR 2015 – Working Track Proceedings, Section no. 5, pp. 1–10, 2015.

[47] P. Gysel, "Ristretto: Hardware-oriented approximation of convolutional neural networks," arXiv preprint arXiv:1605.06402, 2016.

[48] Y. Jia et al., "Caffe: Convolutional architecture for fast feature embedding," in *MM 2014 – Proceedings of the 2014 ACM Conference on Multimedia*, pp. 675–678, 2014.

[49] M. S. Razlighi, M. Imani, F. Koushanfar, and T. Rosing, "LookNN: Neural network with no multiplication," in Design, Automation & Test in Europe Conference & Exhibition (DATE), 2017, Lausanne, pp. 1775–1780, 2017. doi: 10.23919/DATE.2017.7927280

[50] V. N. Vapnik, "An overview of statistical learning theory," *IEEE Transactions on Neural Networks,* vol. 10, no. 5, pp. 988–999, Sept. 1999. doi: 10.1109/72.788640

[51] J. Kivinen, A. J. Smola, and R. C. Williamson, "Online learning with kernels," *IEEE Transactions on Signal Processing,* vol. 52, no. 8, pp. 2165–2176, Aug. 2004. doi: 10.1109/TSP.2004.830991

[52] P. Gastaldo, L. Pinna, L. Seminara, M. Valle, and R. Zunino, "Computational intelligence techniques for tactile sensing systems," *Sensors (Switzerland)*, vol. 14, no. 6, pp. 10952–10976, 2014.

[53] A. Ibrahim and M. Valle, "Real-time embedded machine learning for tensorial tactile data processing," *IEEE Transactions on Circuits and Systems I: Regular Papers,* vol. 65, no. 11, pp. 3897–3906, Nov. 2018.doi: 10.1109/TCSI.2018.2852260

[54] P. Gastaldo, L. Pinna, L. Seminara, M. Valle, and R. Zunino, "A tensor-based pattern-recognition framework for the interpretation of touch modality in artificial skin systems," *IEEE Sensors Journal*, vol. 14, no. 7, pp. 2216–2225, July 2014. doi: 10.1109/JSEN.2014.2320820

[55] J. E. Volder, "The CORDIC trigonometric computing technique," *IRE Transactions on Electronic Computers,* vol. EC–8, no. 3, pp. 330–334, Sept. 1959.doi: 10.1109/TEC.1959.5222693

[56] J. E. Volder, "Birth of CORDIC," *Journal of VLSI Signal Processing Systems for Signal, Image, and Video Technology*, vol. 25.2, pp. 101–105, 2000.

[57] M. Osta, A. Ibrahim, and M. Valle, "FPGA implementation of approximate CORDIC circuits for energy efficient applications," in 2019 26th IEEE International Conference on Electronics, Circuits and Systems (ICECS), Genoa, Italy, pp. 127–128, 2019. doi: 10.1109/ICECS46596.2019.8964758

[58] B. Khurshid and R. N. Mir, "Power efficient implementation of bit-parallel unrolled CORDIC structures for FPGA platforms," in 2015

International Conference on VLSI Systems, Architecture, Technology and Applications (VLSI-SATA), Bangalore, pp. 1–6, 2015. doi: 10.1109/VLSI-SATA.2015.7050466

[59] T. Chang and M. Hsiao, "Carry-select adder using single ripple-carry adder," *Electronics Letters,* vol. 34, no. 22, pp. 2101–2103, 29 Oct. 1998.doi: 10.1049/el:19981706

[60] Z. Yang, A. Jain, J. Liang, J. Han, and F. Lombardi, "Approximate XOR/XNOR-based adders for inexact computing," in 2013 13th IEEE International Conference on Nanotechnology (IEEE-NANO 2013), Beijing, pp. 690–693, 2013. doi: 10.1109/NANO.2013.6720793

[61] C. Liu, J. Han, and F. Lombardi, "A low-power, high-performance approximate multiplier with configurable partial error recovery," in 2014 Design, Automation & Test in Europe Conference & Exhibition (DATE), Dresden, pp. 1–4, 2014. doi: 10.7873/DATE.2014.108

[62] C. I. Allen, D. Langley, and J. C. Lyke, "Inexact computing with approximate adder application," in *IEEE National Aerospace & Electronics Conference Proceedings,* vol. 2015–February, pp. 21–28, 2015.

[63] M. Osta, A. Ibrahim, H. Chible, and M. Valle, "Inexact arithmetic circuits for energy efficient IoT sensors data processing," in 2018 IEEE International Symposium on Circuits and Systems (ISCAS), Florence, pp. 1–4, 2018. doi: 10.1109/ISCAS.2018.8351839

[64] L. Mirsky, F. R. Gantmacher, and K. A. Hirsch, "The Theory of Matrices," American Mathematical Society, Providence Rhode Island, 1959.

[65] V. Klema and A. Laub, "The singular value decomposition: Its computation and some applications," *IEEE Transactions on Automatic Control,* vol. 25, no. 2, pp. 164–176, April 1980. doi: 10.1109/TAC.1980.1102314

[66] A. Ibrahim, M. Valle, L. Noli, and H. Chible, "FPGA implementation of fixed point CORDIC-SVD for E-skin systems," in 2015 11th Conference on Ph.D. Research in Microelectronics and Electronics (PRIME), Glasgow, pp. 318–321, 2015. doi: 10.1109/PRIME.2015.7251399

[67] B. S. Prabakaran, S. Rehman, and M. Shafique, "XBioSiP: A methodology for approximate bio-signal processing at the edge," in 2019 56th ACM/IEEE Design Automation Conference (DAC), Las Vegas, NV, USA, pp. 1–6, 2019.

8

Optical Links for Sensor Data Communication Systems

Andrea De Marcellis*, Elia Palange, Guido Di Patrizio Stanchieri and Marco Faccio

Electronic and Photonic Integrated Circuits and Systems (EPICS) Laboratory, Department of Industrial and Information Engineering and Economics, University of L'Aquila, L'Aquila, Italy
E-mail: andrea.demarcellis@univaq.it; elia.palange@univaq.it; guido.dipatriziostanchieri@graduate.univaq.it; marco.faccio@univaq.it
*Corresponding Author

In recent years, data transmission links based on optical devices demonstrated their capabilities to overcome in many aspects and for specific applications other classical methodologies operating, for example, in the radio-frequency region. When integrated in Si complementary metal–oxide semiconductor (CMOS) technology, optical devices can have size smaller than few micrometers, work in the low-voltage low-current regime, require energies less than few picojoules for each transmitted bit of information even through skin barrier, and operate in the conditions of maximum high electromagnetic compatibility and signal integrity. More in detail, optical communication links make use of a semiconductor laser featuring as the transmitter and a photodiode used as the system receiver. The peculiar medical applications, such as prosthetics devices which we are interested in, require for the coding process the use of light modulation paradigms achieved by acting on the current that drives the laser, data coding schemes to transmit the clock signal and fast electronics for the photodiode conditioning to perform the decoding process. This allows the data transmission from sensor arrays even in implantable biotelemetry systems for the control of prosthetic devices as well as in patients suffering neurological diseases and injuries. The aim of this chapter is to introduce the reader to the working principles of these optical communication systems by describing the coding and decoding procedures that use a combination of optical and analogue/digital electronic architectures designed to be integrated in the Si CMOS technology. As case examples, we will report and discuss two applications concerning sensor data communication in prosthetic systems and neural implantable devices.

8.1 Introduction

The future devices for industrial, medical, and life sciences applications will employ an increasing number of sensors to measure different physical and chemical parameters such as pH values, humidity, temperature, spatial parameters, object shapes, and surface roughness. The data generated by the sensors must be acquired and elaborated to perform autonomous operations of equipments and vehicles. For example, the improvements in biomedical engineering allow extraction and processing of information carried out by neural and biological signals, and to control external electrics and electronics life-aid apparatus, such as body-machine- and/or brain–computer interfaces, with the aim to ensure a satisfactory life quality of patients with physical and/or neurological diseases [1–6]. In these regards, prosthetic limb, for example, must be equipped with arrays of tactile sensory systems interacting with the external environment to restore as much as possible the sense of touch of a human limb. In this sense, the complete sensory system must be able to return the most important features of the touched objects such as their shape, size, surface roughness, temperature, consistency, and weight. On the other hand, detection, recording, and stimulation of cellular and neuronal signals and/or activities allow recognizing neural coding processes [7–14]. Also, for the neuronal clinical applications a large number of sensors is necessary to record these signals that can come-from or go-to the brain cortical area by means of the design and implementation of implantable bidirectional biotelemetry links that connect specific internal parts of the patient to be monitored or activated by external equipment. The previous two applications are examples, similar to many others, that highlight the aim of the research in neuronal medical fields that, for many aspects, are not so dissimilar from those ones related to the developments of humanoid robots. All these applications have in common the acquisition of a large number of data from different kinds of sensors, the elaboration of these data to permit autonomous decision-making activities with the subsequent elaboration of procedures for the activation of devices and/or the generation of stimuli. In the case of the movement of prosthetic devices, when, for instance, mechanical stresses are applied on the touched object or temperature variations occurred, corresponding electrical signals are generated by the sensor arrays. At that time, these analogue signals must be transmitted from the sensors to a read-out circuitry that uses them as the input data. The circuitry, in turn, provides the signal digital conversion by using an analogue-to-digital converter (ADC) and then, passes the digitized data to a processing unit. In general, the processing unit runs a statistical learning algorithm on the data to extract the meaningful information about the physical variations of the touched object and, on this basis, drives a stimulator that provides to generate a series of electrotactile stimuli understandable by the user [15–17]. It is worth noting that a large number of sensors (i.e., tactile sensors together with other types of sensors depending on the specific application and use) must be employed with fast response electronic circuitries to obtain human-like touch sensing capabilities [18–20]. Similar considerations can be done for the transcutaneous implanted systems that acquire neural signals and transfers them from inside to outside of the patient (and vice versa) suffering physical and/or neurological diseases. The neural signals are analogue electrical pulses detected by micro- and nanosensor arrays directly inserted into the patient's brain [21–26]. An implanted electronic circuitry is used to acquire the signals from these sensors and to digitally decode them in a form to be transmitted to external controlling and/or actuation devices. In this case, whatever be the transmission methodology, the implemented transcutaneous telemetries must operate at high data rates with exceptional energy efficiency in terms of very low overall power consumption. This is an important system constraint for ensuring the health of the patient since the power spent to activate the telemetry operations for the data acquisition and transmission is dissipated as heat in the tissue and can cause severe damage if the dermal thermal limits are exceeded. For this, the consequent specific requirement in transcutaneous implanted systems is to obtain their operation in the low-voltage and low-current regime [27–33]. From the above discussed applications, it would be clear that the type of sensors to be employed is directly related to the specific applications but the general requirements are the design and fabrication of electronic analogue/digital circuitries for the acquisition, elaboration, and transmission of a very large number of data with fast response times (i.e., very large frequency bandwidth) to compel the requirements of an immediate response to stimuli and the activation of body-machines, prosthetics devices, and/or brain–computer interfaces. In this chapter we will discuss the details of the implementation of the optoelectronic systems together with the electronic analogue and digital circuitries designed for the coding and decoding processes of the sensor-generated voltage signals for the two applications outlined above: activation, control, and transmission of stimuli for tactile sensory feedback in prosthetic devices, and implantable biotelemetry system for neural signal recording and stimulation [34–49]. For both these two applications the transmission of the coded data toward the decoding circuitry is achieved by means of an optical link that is composed of a large bandwidth semiconductor laser and Si photodiode (PD). For the coded data transmission can be used other different approaches such as simple wire connections in percutaneous systems that are uncomfortable and potentially cause infections and diseases [50, 51] and radiofrequency techniques employing antennas as the wide band transmitters and receivers [52–54]. The drawbacks of using these approaches for the applications which we are interested in, are the resulting poor electromagnetic compatibility and signal integrity [55, 56] and the increase of the electrical power as the operation frequency increases that can generate dermal injuries. Moreover, these data transmission techniques cannot be or are difficult to be integrated in small dimensions even if attempts have been recently reported in minia-turizing the radiofrequency devices [57–59]. On the other hand, the optical data transmission links based on optoelectronic components avoids all these drawbacks because they are ideally insensitive to electromagnetic distur-bances (i.e., the optical carrier frequencies are orders of magnitude higher than those ones of the electromagnetic radiation achievable with electronic circuitries), guarantee the best achievable signal integrity with minimum bit error rate (BER), can be fabricated using the standard Si CMOS integrated technology in micrometer square resulting area so as to allow operations at very low-voltage and low power. Moreover, as it happens in the optical long/medium/short haul telecommunication networks, the achievable very large frequency bandwidth of the optoelectronic devices (i.e., laser and PD) and of the optical fibers used for the optical links allows using the same transmission channel to transmit the signals generated by different kinds of sensor arrays. This is an important advantage of the optical links with respect to other possible transmission technologies and greatly simplifies the hardware of the prosthetic and biomedical devices. In relation to the specific application, the optical transmission links can work in both the free space wave propagation and guided method by using optical fiber connections. The latter is preferred to transmit data in prosthetic devices and in robots, while the free space propagation is used in implantable system to avoid using wired and radiofrequencies techniques [60–63]. As it will be detailed in the next section, the typical optical transmission link requires to accomplish the following implementation steps: (i) the ADC conversion of the sensor signals and the data preprocessing [64–66]; (ii) the coding process for the generation of a sequence of digital data by using a suitable coding scheme for the transmission of *Clock* synchronization signal also [67–75]; (iii) the generation of a sequence of current pulses replica of the digital ones; (iv) the generation of a sequence of laser pulses replica of the current ones; (v) the conversion of the sequence of laser pulses in a sequence of voltage pulses generated by the PD; and (vi) the transmission of the *Clock* recovery for the final decoding process and data postprocessing. In the following sections all of these steps will be described in detail for tactile sensory feedback systems in prosthetics devices and for implantable biotelemetry systems.

The chapter is organized as follows: in Section 8.2 we will describe the principles of operation of the optical communication systems for sensor data transmission; Section 8.3 is devoted to present some technical solutions for the implementations of the optical links. In particular, we will present a description of: (i) the digital architectures for the coding and decoding processes of the sensor data and (ii) the analogue circuits for the sensor signal conditioning. The proposed circuit solutions have been designed to be implemented by using commercial apparatus and/or components but it will be demonstrated that their architectures allow the circuitries to be directly integrated at the transistor level by using the Si CMOS technology; in Section 8.4 we will present the applications of the optical communications systems for the actuation of prosthetic devices and for the implementation of implantable biotelemetry systems. Finally, our conclusions will be presented to highlight the major contributions of the application of the optical communication links in the general field of biomedical telemetry.

8.2 The Optical Communication Link: Principles, Data Coding, Architectures, and Devices

The understanding of the general methodology to be used for the design and implementation of optical communication links for biotelemetry applications is facilitated by analyzing the transmitter and receiver modules that are schematically presented in Figure 8.1.

In particular, the transmitter module is composed of two blocks and of a vertical cavity surface emitting laser (VCSEL). DATA CODING performs the coding of the data that have been previously digitally converted by

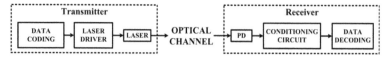

Figure 8.1 The transmitter and receiver modules of an optical communication link.

using an ADC. The output of this block is a sequence of voltage pulses suitably codified that must also contain a synchronization *Clock* signal needed to properly transmit and receive the information contained in the signal sequence generated by the sensors. The second block is the LASER DRIVER that receives the sequence of the coded pulses and transforms it in a sequence of current pulses, exactly a replica. This sequence of current pulses must have an amplitude greater than the threshold level to activate the VCSEL laser action (i.e., the amplitude of each current pulse of the sequence must exceed the value of the VCSEL threshold current). In this way, the VCSEL generates a sequence of laser pulses that is the counterpart of that one generated by the coding process. Thus, the digitally converted data containing the information on the changes of the physical/chemical parameters measured by the sensors are optically transmitted. As outlined in the introduction, the generated laser pulses can propagate in the free space or be coupled to single or multimode optical fibers. It is important to note that VCSELs are the light sources commonly chosen for the optical biotelemetry communication systems because they have a physical structure compatible with the layer-by-layer fabrication procedure of the state-of-the-art Si CMOS technologies that are capable to integrate on the same Si chip electronic and photonic devices [76, 77]. In any case, even if a direct integration is not possible, with respect to the longitudinal emitting semiconductor lasers, the particular VCSEL structure allows for their postproduction integration on the Si chip. From Figure 8.1, we also observe that the receiver module is composed of two blocks and of an optoelectronic device, the PD. The PD, usually a Si photodiode with frequency bandwidth equal or larger than that of the VCSEL, generates current pulses that follow the same temporal shape of the transmitted laser pulses with amplitudes proportional to their intensities. By using an electrical scheme that employs a transimpedance amplifier (TIA), the output of the conditioning circuit is a sequence of voltage pulses from which it is possible to obtain the information transmitted by the optical communication link through a suitable decoding operation. It is worth mentioning that the decoding procedure can be properly executed only if the transmitted *Clock* signal is correctly recovered. Figure 8.2 helps to understand how the transmitted sequence of laser pulses also contains the *Clock* signal.

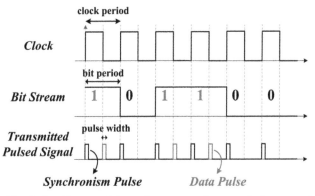

Figure 8.2   The timing diagram of the coding process for an optical communication link.

Starting from the top of Figure 8.2, the first periodic waveform is the *Clock* signal. For each *Clock* period a bit {1} or {0} of the *bit stream* must be transmitted. The *Clock* signal is recognized by the decoding block because for each *Clock* period a laser pulse is generated independently from which bit is transmitted. Thus, this represents the *synchronism pulse* having the same period of the *Clock* signal and is used for the *Clock* recovery operation.

Referring now to the *bit stream* waveform of Figure 8.2, for this timing architecture, if a bit {1} must be transmitted, a laser pulse is generated at a time corresponding to a half-period of the *Clock* signal. On the other hand, if a bit {0} must be transmitted, no laser pulse occurs at a time corresponding to a half-period of the *Clock* signal. Thus, the resulting *transmitted pulsed signal* is formed by an aperiodic sequence of laser pulses where only those transmitting the *Clock* signal have a defined frequency. The described transmission paradigm is a modified version of the classical OOK modulation scheme and has been designed for the systems that must operate in the low-voltage and low-power regime. With respect to the OOK modulation scheme, in fact, Figure 8.2 shows that the laser operates only for a time corresponding to the laser pulse duration that can be shorter or much shorter than the *Clock* period. As a consequence, this operation mode minimizes the power consumption of the overall optical transmission link.

8.3 Technical Solutions and Implementations of Optical Links

The technical solutions described in the following paragraphs, consider that the optical transmission links can operate with the laser pulses propagating in the free space or in a guided mode. The first operation corresponds to the optical wireless transmission mode while in the second one the optical transmission link makes use of optical fibers in which the laser beam is properly coupled. The optical wireless transmission technique is employed in all those applications for which a connection by an optical fiber must be avoided as in the case of human implantable biotelemetry links. On the other hand, the optical fiber connections are preferred in those applications for which the laser beam must overcome physical obstacles or propagate in a nonstraight direction as in the prosthetic and humanoid robotic applications. Moreover, the presented electronic and optoelectronic architectures have been designed to achieve efficient and reliable optical transmission links and have been firstly implemented by using discrete components on printed circuit boards (PCBs) for the analogue electronic circuitries and field programmable gate array (FPGA) boards for the digital ones. However, all the technological solutions have been conceived by considering that the chosen solutions can be easily adapted for the fabrication of the entire optical transmission link in a single chip by using the standard Si CMOS integrated technology.

8.3.1 Description of the Digital Architectures for the Coding and Decoding Processes of the Sensor Data

The implementation of the data coding technique schematically discussed in the previous section requires the design and the development of suitable digital architectures. The possible solutions must ensure reliable coding and decoding operations even at high data rate with overall high power efficiency [78]. For these purposes, an FPGA-based architecture has been used for the preliminary implementation of the digital solutions and for the fast prototyping of the system using commercial discrete components. In particular, the following described architectures have been implemented on a Xilinx FPGA Kintex Ultrascale board but the proposed digital solutions can be implemented also on other FPGA board families. According to Figure 8.3, the data coding can be implemented by using a phase-locked loop (PLL) and two logic gates. The PLL, already realized as a basic block inside the FPGA, generates two pulsed signals starting from the input *Clock* signal (i.e., the *synchronism pulse*). The first pulse at the PLL terminal A is generated in correspondence of the rising edges of the *Clock* signal, and the second pulse at the PLL terminal B is generated synchronized with the falling edges of the *Clock* signal. These two pulsed signals have the same frequency with a relative phase difference of 180° and a selectable duty cycle to guarantee the

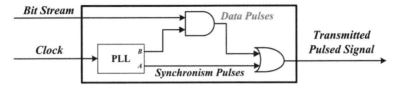

Figure 8.3 Digital logic scheme of the DATA CODING block.

desired pulse width. The signal at the PLL terminal B is combined with the *bit stream* (Figure 8.2) through an AND digital logic gate so that the *data pulses* are generated only when the *bit stream* assumes the "high" logic state (i.e., corresponding to the logic value {1}). Finally, the output of the AND gate is combined with the *synchronism pulses* by means of an OR digital logic gate. Consequently, the resulting combination of these two pulsed signals provides a train of voltage pulses corresponding to the *transmitted pulsed signal* that is employed as the input control signal of the LASER DRIVER block of Figure 8.1. In Figure 8.4 is reported the digital logic scheme of the DATA DECODING block. Starting from the *received pulsed signal*, the *Clock* recovery sub-block recovers and regenerates the *Clock* signal needed for the data synchronization. Simultaneously, the IDELAYE3 primitive block processes the same *received pulsed signal* to start the data recovery procedure. This is a programmable time delay line implemented into the I/O blocks of the FPGA that provides a finite and discrete time delay to be added to the input pulsed signal. As a consequence, the IDDR primitive block allows for the *recovered bit stream* that is the recovered pulse/data present at the falling edge of the *recovered clock* starting from the properly delayed *received pulsed signal* received at its input.

At the starting time of the DATA DECODING block, the control unit DECOD (C.U. DECOD) gradually increases the time delay introduced by IDELAYE3 until the rising edge of the *recovered clock* is in-phase (i.e., synchronous) with the *synchronism pulses*. In this way, the falling edge of the *recovered clock* allows to recover the *bit stream* from the received *data pulses*. Furthermore, to perform a compensation of the time delay variations of the IDELAYE3 due to supply voltage and/or operating temperature drifts, the C.U. DECOD properly enables and controls also the IDELAYCTRL block that is a further primitive entity used for this specific purpose implemented on the FPGA. As shown in Figure 8.5, the *Clock* recovery block is is simply composed by a flip-flop (FF), a PLL, and a latch block. By means of the latch block a specific time delay is introduced to reset the FF that occurs between the *data pulse* transition and the following *synchronism pulse*. In this

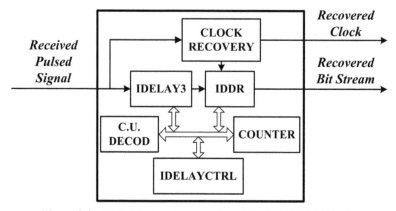

Figure 8.4   Digital logic scheme of the DATA DECODING block.

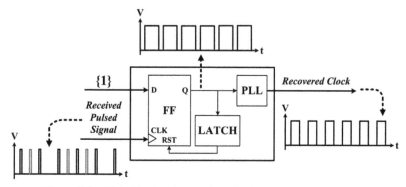

Figure 8.5   Digital logic scheme of the CLOCK RECOVERY block.

way, the FF results to be only sensitive to the *synchronism pulse* rising edges. Consequently, the regenerated *Clock* signal, provided by the FF at the output Q, has a duty cycle higher than 50%. In these conditions, the PLL performs a duty-cycle correction so that the *recovered clock* signal has a duty cycle equal to 50%. More in detail, the input D of the FF is always forced in the "high" logic state (i.e., the bit {1}) while the *received pulsed signal* represents its input *Clock* signal. Thus, when a pulse of the *synchronism pulses* occurs, the output Q of the FF that was initially reset reaches the "high" logic level and remains in this status until the FF is again reset. The output Q of the FF acts on a feedback loop including the latch block that controls and allows to reset the FF.

In particular, the output Q remains in the high logic state for a time duration $T_{FF} = T_{Route} + T_{Latch} + T_{RST}$, where T_{Route} is the physical time

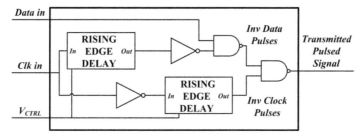

Figure 8.6 The architecture at block scheme level of the DATA CODING block.

delay introduced by the route that connects the FF output Q to the latch block and this one to the FF RST input, T_{Latch} is the response time of the latch block and T_{RST} is the time necessary for the FF to make effective the reset operation. Moreover, the output of the PLL keeps always the *recovered clock* signal locked even in the case of some missing pulses, in consequence of transmission errors. Once properly asserted the functionality of the digital coding and decoding architectures implemented by discrete components, it is possible to integrate them on a chip using the standard Si CMOS technology maintaining the same system functionalities [79]. Moreover, a full custom design avoids the use of elements such as a PLL and reduces the power consumption and circuit complexity of the overall system. In the following, we will discuss different microelectronics solutions implemented by standard Si CMOS technology. Referring to Figure 8.6, the DATA CODING architecture is based on simple combinational logic blocks, consisting of $2\times$ NOT, $2\times$ NAND, and $2\times$ rising edge triggered pulse generators (i.e., the RISING EDGE DELAY blocks in Figure 8.6). The schematic circuits of these elements are reported in Figure 8.7(b), Figure 8.7(c), and Figure 8.7(a), respectively. In Figure 8.7(a), starting from the *Clock* input *Clk in*, the RISING EDGE DELAY block generates a sequence of voltage pulses with a variable width regulated by the control voltage V_{CTRL}. This control voltage is based on three inverter stages formed by the transistor pairs M1–M2, M11–M12, and M14–M15 and operates as follows: when the voltage $V_{CTRL} = 0$ V, the diode-connected transistor M5 defines the maximum resistance in the pull-down network of the M1–M2 inverter stages. In a similar way, the diode-connected transistor M10 sets the maximum resistance in the pull-up network of the M11–M12 inverter stage. Under these operating conditions, the maximum time delay is achieved. By increasing the value of V_{CTRL}, the transistor M3 and current mirrors M6–M7–M9 progressively turn on, and increase the current flowing through the transistors M4–M8. This results in an

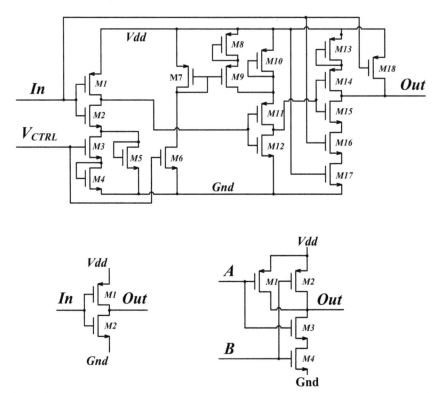

Figure 8.7   Example of schematic circuits at CMOS transistor level: (a) RISING EDGE DELAY block; (b) NOT digital logic gate; and (c) NAND digital logic gate.

overall reduction of the delay between the rising edges of the input and output signals. The circuit, however, is designed to directly respond to the falling edge of the input *Clock* signal. This is achieved through the transistor M18 in such a way that only the input rising edge is delayed by the RISING EDGE DELAY block. Referring now to the entire digital coding block of Figure 8.6, the input *Clk in* follows two different paths to provide voltage pulse trains with a relative phase shift equal to 180° (i.e., equal to the half-*Clock* period).

These signals are then combined with the *data in* input data stream through the NAND gates to generate a single output pulse train that contains both the *Clock pulses* and the *data pulses* signals. The DATA DECODING system presented in Figure 8.8 is able to perform the data and the *Clock* recovery. The data decoding process is achieved by using 2× D-type FF (FF1 and FF2 specified in Figure 8.9[b]), 2× inverters, 1× RISING EDGE DELAY block (equal to that one described earlier) and 1× PHASE CONTROL block.

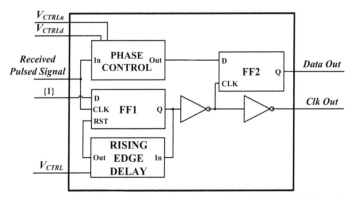

Figure 8.8 The architecture at block scheme level of the DATA DECODING block.

According to Figure 8.9(a), The DATA DECODING operates as follows: The *received pulsed signal* drives the *Clock* input CLK of FF1 with the data input D fixed to the logic level {1}. In this way, the first incoming *synchronism pulse* sets the output of the FF1 to a high logic level and thus, any further *data pulse* cannot affect the FF1 output. The FF1 output signal is then maintained through the RISING EDGE DELAY block. After this fixed time delay, the RISING EDGE DELAY triggers the FF1 asynchronous reset pin RST to toggle its output back to the low logic level {0}. The system is now ready to accept the next *synchronism pulse*. Since there is a fixed phase relationship between the *synchronism pulses* and the *data pulses* signals, it is essential to adjust the pulse delay for different data rates through the RISING EDGE DELAY block to guarantee a delay value between T/2 and T, with T the *Clock* period (i.e., the period between two consecutive *synchronism pulses* of Figure 8.2).

Consequently, the output of FF1 provides the recovered *Clock* signal *Clk Out* but with a duty cycle higher than 50% (i.e., half of a *Clock* period plus the additional pulse delay). Thus, FF2 receives the inverted recovered *Clock* and acquires the data input provided by *data pulses*. To avoid a metastable state in FF2, a PHASE CONTROL has been included to guarantee the FF2 setup and hold times. Referring to Figure 8.9(a), the PHASE CONTROL block consists of 4× inverter stages implemented by the transistor pairs M1–M2, M5–M6, M7–M8, and M9–M10. The transistors M3–M4 are also driven by the control voltages V_{CTRL_U} and V_{CTRL_D} that allow for the pulse width of the *received pulsed signal* to be extended. Finally, FF2 provides the decoded data *data out* signal with a stable value corresponding to the rising edges of the recovered *Clock* signal *Clk Out*.

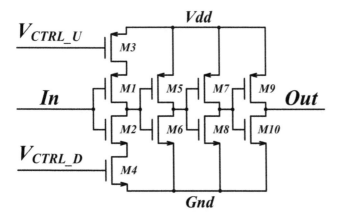

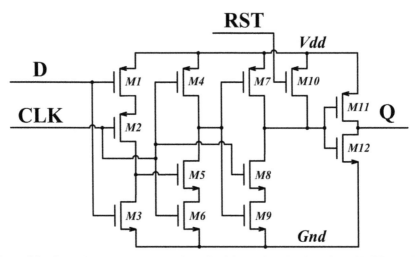

Figure 8.9 Example of schematic circuits at CMOS transistor level: (a) PHASE CONTROL block and (b) FF block.

8.3.2 Description of the Analogue Circuits for Sensor Signal Conditioning

In this section we will describe some possible solutions for the implementation of the analogue electronic circuitries for the signal conditioning in optical communication links. Preliminary implementations and fast prototyping of the designed circuits have been obtained by using commercial discrete components. In particular, referring to the transmitter module of the system, the

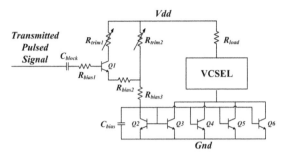

Figure 8.10   Example of the schematic circuit of the LASER DRIVER block.

VCSEL can be controlled through a driving circuit (i.e., the LASER DRIVER block of Figure 8.1) based on current mirrors, as shown in Figure 8.10. This circuit topology allows the conversion of the voltage pulses (i.e., the *transmitted pulsed signal*) in current pulses to drive the VCSEL. The circuit presents also the capability to regulate both the pulsed current amplitude and a DC current level through two resistive trimmers R_{trim1} and R_{trim2}, respectively. The laser action takes place if the sum of the pulsed and DC currents is greater than the VCSEL threshold current. In these conditions, the resulting laser pulse peak intensity is proportional to the value of the corresponding driving current pulse amplitude.

For what concerns the receiver module, Figure 8.11 shows an example of the schematic of the PD CONDITIONING CIRCUIT block (see also Figure 8.1) that employs a multiple-stage topology of TIA. It is based on a cascade of Darlington architectures converting the pulsed photocurrent (i.e., the generated photocurrent pulses) into voltage pulses (i.e., the *received pulsed signals*) having an amplitude higher than the digital logic threshold needed for a correct detection of a "high" logic level (i.e., the bit {1}) by the subsequent DATA DECODING block. These analogue circuits can be easily implemented on prototype PCBs with discrete off-the-shelf components (e.g., high frequency operations transistors for RF applications) operating also at low supply voltage (e.g., 3.3 V). On the other hand, considering the possibility to develop integrated solution in a standard CMOS technology, in the following are reported examples of possible integrated solutions designed at the transistor level. In this regard, a simple solution of a LASER DRIVER circuit is reported in Figure 8.12. The circuit is based on a current mirror stage formed by the transistors M4–M5, converting the input voltage pulses (i.e., the *transmitted pulsed signal*) into current pulses to directly drive the VCSEL. Moreover, as discussed earlier, this allows for the regulation of

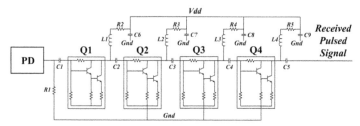

Figure 8.11   Example of the schematic circuit of the PD CONDITIONING CIRCUIT block.

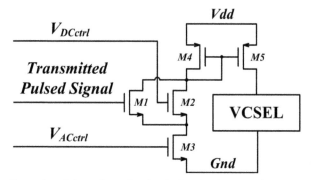

Figure 8.12   Example of the schematic circuit for the integrated solution of a LASER DRIVER block.

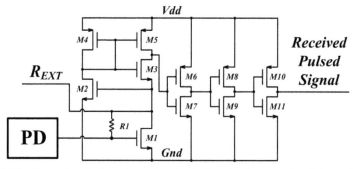

Figure 8.13   Example of the schematic circuit for the integrated solution of the PD CONDITIONING CIRCUIT block.

both the pulsed current amplitude and the DC current level through the two control voltages V_{DCctrl} and V_{ACctrl} that act on the transistors M2 and M3, respectively.

Referring now to the receiver module, an example of the schematic of the PD CONDITIONING CIRCUIT block is reported in Figure 8.13. The circuit is based on a single stage TIA implemented by the transistors M1–M5, with the resistor R1 controlling the transimpedance gain. Additional gain is provided through a cascade of three CMOS inverter stages implemented by the transistor pairs M6–M7, M8–M9, and M10–M11 that generate the output signal *received pulsed signal*. Furthermore, the R_{EXT} terminal allows for the overall gain to be precisely adjusted through an optional external resistor connected in parallel with R1.

8.4 Examples of Applications of Optical Communication Links for Sensory Systems

In this section we will present the applications of the optical communications systems for the actuation of prosthetic devices and for the implementation of implantable biotelemetry systems.

8.4.1 Optical Fiber Link for Prosthetics Developed by Discrete Commercial Components and Devices

Figure 8.14 shows the implementation of the optical communication link, based on optical fiber, for sensory feedback in prosthetic devices. The system consists of an array of tactile sensors placed on the palm of the prosthetic hand. The generated tactile sensor signals are first ADC converted and thereafter digitally processed and optically transmitted by the optoelectronic circuitry present in the forehand. Once received and processed, the transmitted signals are used to activate an electrocutaneous stimulator in contact

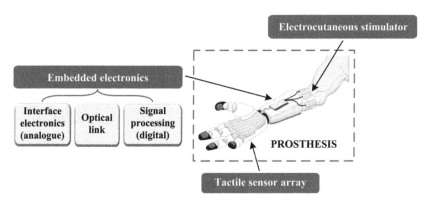

Figure 8.14   Example of a data communication system for tactile sensory feedback in prosthetic devices.

with the arm of the patient to restore the sense of touch obtained through the tactile sensors. It is worth noting that this represents a first, simple example of application of the optical communication link where the number of data to be transmitted is not so large.

The strategic utility of optical communication architecture is more evident if the number and type of the sensors (i.e., temperature, humidity, shape, etc.) located in the prosthesis increases, the prosthesis from passive becomes active with the use of a number of electrical motors that must be activated to simulate a real human hand. In this case, the very large bandwidth of the employed optoelectronics devices will be able to transport all the signals within a single optical communication link. The implementation scheme of the system is presented in Figure 8.15 [62, 63]. The solution employs two different FPGAs (Spartan6 and Virtex6 by Xilinx) to implement the transmitter and the receiver modules. The optical communication link is provided by a VCSEL (OPV314AT by TT Electronics), a PD (DET025AFC/M by Thorlabs), and an optical fiber (50/125 μm multimode optical fiber). Both the laser beam and the PD sensitive area are directly coupled to the fiber by using two lenses with the appropriate numerical apertures to minimize the light power losses. The LASER DRIVER circuit (see its schematic in Figure 8.10) employs the BFP720 low noise RF Si–Ge bipolar transistors and the PD CONDITIONING CIRCUIT (see its schematic in Figure 8.11) is implemented through a cascade of four ERA-1SM+ InGaP HBT wideband monolithic Darlington pair to achieve very high gain values. These analogue circuitries have been developed using the PCBs paradigm by employing a TLX8 substrate and the above mentioned commercial off-the-shelf components. More in detail, referring to Figure 8.15, after the *start* signal has been set to the logic state equal to {1}, the ADC INTERFACE block is enabled to generate a proper *Clock ADC* signal to retrieve the data from an external data acquisition module operating at few kS/s conversion rate (i.e., 2kS/s by employing a DDC232 20 bit 32 channels ADC by Texas Instruments). Every time the TX MODULE toggles the signal *Start Conv*, the external ADC simultaneously scans and converts the analogue signals generated by the array of tactile sensors. The converted signals are shifted out to the acquisition module through a serial interface/protocol. Thus, when the acquisition and the conversion is accomplished, the *data valid* pin of the ADC goes low indicating that the data are ready to be stored in the internal global data buffer. Starting from the acquired data, the ADC INTERFACE block generates the serial data package to be transmitted. In this way, according to Figure 8.16, the serial data package consists of a well-ordered sequence

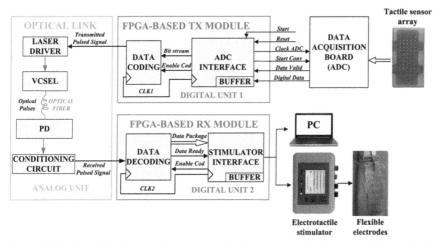

Figure 8.15 Overall implementation scheme of a data communication system for prosthetic devices.

of samples (each one corresponding to the related tactile sensor of the input array) and a header used to detect the beginning of the package.

Subsequently, the serial *data package* is transmitted at a data rate of 100 Mbps using the coding circuits and the architecture solutions described in Section 8.3. Once the package is regenerated by the DATA DECODING block, the data can be stored into the global data BUFFER, suitably processed, and sent to a stimulator and/or to a PC monitor through a standard UART communication protocol implemented by the UART INTERFACE block or simply evaluated by an oscilloscope. In particular, the data are processed to provide proper control commands to the stimulator device together with the generation of stimuli corresponding to the touch detected by the input tactile sensors. The control commands carry out all the parameters related to the stimulations to be generated, such as the electrostimulation pulse intensity, the frequency, and the electrode channel position. These parameters can change according to the type and the force intensity of the touch of the sensing elements (i.e., their physical stimulation). Once the data BUFFER is empty, another *data package* can be acquired, stored, and processed. In Figure 8.17 is presented an example of the measurements performed by employing the implemented system demonstrating that the receiver is able to perform the data and the *Clock* recovery starting from the *received pulsed signal*. The detection of the package through a HEADER identifies the beginning of the transmitted data acquired from the array of sensors.

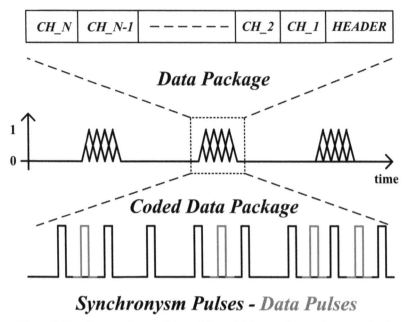

Figure 8.16   Structure/composition of the serial data package to be transmitted.

8.4.2 Optical Wireless Communication Integrated System for Implanted Biotelemetry Applications

Figure 8.18 shows the overall architecture of an integrated transcutaneous optical wireless bidirectional biotelemetry link implementing the UWB-inspired pulse data coding based on a synchronized-OOK modulation as described in Section 8.2. The complete system includes a transmitter and a receiver within each one of the implanted and external modules separated by a tissue/skin layer and allows for a high bit rate link with low power consumption.

As the input signals, the transmitter takes the main *Clock Clk In* and the bit stream *Data In* to be coded and transmitted. The bit stream can be generated, for example, by a neural recording apparatus. On the other hand, the receiver provides the recovered *Clock* signal *Clk Out* and the decoded bit stream *data out* to be sent to external monitoring devices and/or actuators. These two subsystems operate independently and can work also at different operating frequencies thus allowing for up- and downlinks at different bit rates. In particular, the TRANSMITTER module includes the DATA CODING block to modulate the voltage pulses and a LASER DRIVER block for biasing and

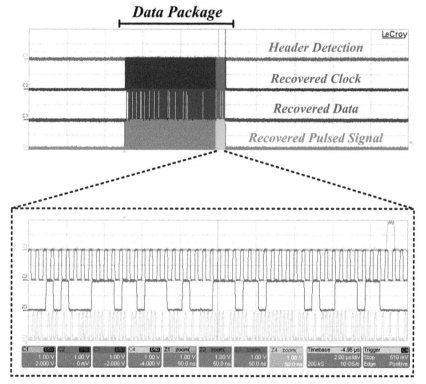

Figure 8.17 Example of the experimental measurement on the implemented communication system for prosthetic devices.

driving the VCSEL by converting the voltage pulses into current pulses to generate subnanosecond laser pulses. The resulting output signal from the digital blocks provides the coded voltage pulses (i.e., the *transmitted pulsed signal*) that are converted into current pulses by the VCSEL driving circuit (i.e., the LASER DRIVER) to generate the laser pulses. The receiver includes a CONDITIONING CIRCUIT that converts the current pulses (generated by a fast Si PD) into voltage pulses and a DATA DECODING block that takes the received voltage pulses and performs the *Clock* and data recovery. The overall system has been designed and fabricated in 0.35 μm CMOS technology (AMS C35B4C3 process) as a single-chip full-custom ASIC in a very compact Si footprint including all the components except the VCSEL and the PD that have been mounted externally. The core system occupies approximately 0.13 mm^2. In particular, the transmitter is composed of 53 transistors and 1 resistor while the receiver contains 60 transistors (see

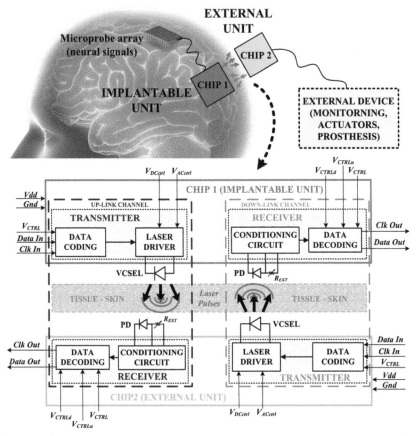

Figure 8.18 System overview of a transcutaneous optical bidirectional biotelemetry link.

Section 8.3). All the circuits operate using a single 3.3 V supply voltage. The design has been optimized for high data throughput, low power consumption, and small Si area. Moreover, all the blocks that have been described earlier have been designed targeting a full-duplex link for a bidirectional communication paradigm with up- and down-link transmissions at bit rates up to 250 Mbps for each communication channel. The microphotograph of the fabricated ASIC including the pad ring (i.e., ESD protection I/Os and physical bond pads) is shown in Figure 8.19 with the main blocks highlighted in red. The overall chip (being a pad-limited design) measures 1.6×2 mm^2 and has been encapsulated into a JLCC44 (44 pins) ceramic package. The chip has been evaluated through a custom-designed PCB that houses the ASIC and allows for connection to external modules and instruments for

**CONDITIONING
CIRCUIT**

DATA
CODING

DATA
DECODING

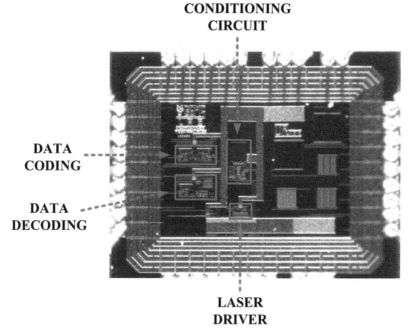

**LASER
DRIVER**

Figure 8.19 Microphotograph of the fabricated ASIC where the main modules are identified within the red line boxes.

the test measurements. The complete system is capable to correctly transmit and receive data operating also in a full-duplex mode. It employs VCSELs (VCSEL-850 by Thorlabs) emitting at a wavelength of $\lambda = 850$ nm and high-speed PDs (FDS-025 by Thorlabs). Additionally, a suitable test bench has been implemented on a Xilinx VIRTEX-6 XC6VLX240T FPGA board for the generation of the bit streams, the master *Clocks*, and the pulse trains [80–82]. A photo of the experimental apparatus is shown in Figure 8.20. According to Figure 8.18, two ASIC have been mounted facing each other to implement a simultaneous bidirectional link by using two couples of VCSEL and PD.

Two XYZ TRANSLATION STAGEs allowed for an accurate optical alignment of the VCSEL and the PD along the X- and Y-directions and for the regulation of their relative distance along the Z-direction (i.e., the laser beam propagation direction perpendicular to the XY plane). Moreover, two 1.5 mm DIFFUSERs ED1-C20-MD (Thorlabs) have been inserted between the VCSEL and the PD to emulate the skin/tissue effects such as light attenuation, diffusion, and scattering. Figure 8.21 shows an example of the measured

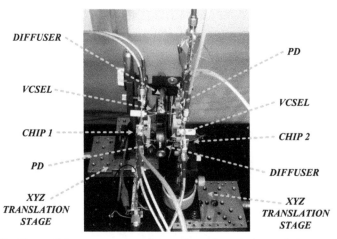

Figure 8.20   Photo of the setup employed to perform the experimental characterizations of the fabricated integrated biotelemetry system.

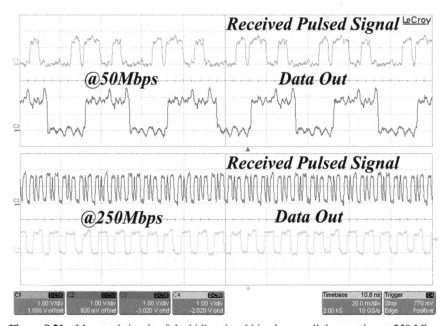

Figure 8.21   Measured signals of the bidirectional biotelemetry link operating at 250 Mbps (uplink channel) and 50 Mbps (downlink channel) transmitting a $\{0, 1\}$ bit sequence.

signals illustrating the encoding, the levels, and the timing throughout of the system working correctly in the full-duplex mode.

The coded temporal pulse width is about 1.2 ns with a maximum peak current level driving the VCSEL of approximately 24 mA. The transmitted bit stream used is composed of a repetition of $\{0, 1\}$ bit sequences to quickly and better evaluate the correctness of the system operation and functionality operating at different bit rates. In this case, the uplink channel transmission bit rate is 250 Mbps while the downlink channel bit rate is equal to 50 Mbps. Finally, the complete system implemented onto the CHIP 1 (i.e., the implantable unit) requires a maximum total power consumption of about 38 mW with both the transmitter and receiver modules active, thus achieving a total power efficiency of about 214 pJ/bit. These data have been obtained with the best optical alignment along the X- and Y-directions and a maximum distance of 3 mm between the VCSEL and the PD along the Z-direction. This distance includes the diffuser that reduces the laser power by a factor of 10.

8.5 Conclusion

The aim of the chapter was to introduce the reader to the design and implementation of optical communication links for the transmission of sensor and neural data in prosthetic devices and in implantable biotelemetry systems. We started from the description of how the voltage pulses generated by different kinds of sensors can be optically transmitted by using optoelectronic devices. Then, we have reported the design and the implementation of specific analogue/digital circuitries and data coding and data decoding architectures. All the proposed circuit solutions can be implemented and developed by using commercial apparatus and components, but we have also demonstrated that the used paradigms allow for a direct integration of all the circuits at the transistor level by using standard Si CMOS technologies. Two practical examples of applications have been presented to prove the capability of the optical communication links to be used for the data transmission in tactile sensory feedback systems in prosthetic devices as well as for bidirectional transmission (i.e., in the down- and up-link conditions) of neural data in implantable biotelemetry systems. The experimental findings indicate that the optical communication links guarantee very high values of electromagnetic compatibility and signal integrity, capabilities to operate in low-voltage low-power regime, transmission data rates up to 250 Mbps, and overall power efficiencies of few hundreds of pJ/bit.

References

[1] R. Carvalho, N. Dias, J.J. Cerqueira, 'Brain-machine interface of upper limb recovery in stroke patient rehabilitation: A systematic review', Physiotherapy Research International, 24, e1764, 2019.

[2] J. Ribeiro, F. Mota, T. Cavalcante, I. Nogueira, V. Gondim, V. Albuquerque, A. Alexandria, 'Analysis of man-machine interfaces in upper-limb prosthesis: A review', Robotics 8, p. 16, 2019.

[3] M.W. Slutzky, 'Brain-machine interfaces: Powerful tools for clinical treatment and neuroscientific investigations', The Neuroscientist, 25, pp. 139–154, 2019.

[4] A.N. Belkacem, S. Nishio, T. Suzuki, H. Ishiguro, M. Hirata, 'Neuromagnetic decoding of simultaneous bilateral hand movements for multidimensional brain–machine interfaces', IEEE Transactions on Neural Systems and Rehabilitation Engineering, 26, pp. 1301–1310, 2018.

[5] R.P.N. Rao, 'Towards neural co-processors for the brain: combining decoding and encoding in brain–computer interfaces', Current Opinion in Neurobiology, 55, pp. 142–151, 2019.

[6] S. Shahdoost, P. Mohseni, 'An FPGA platform for generation of stimulus triggering based on intracortical spike activity in brain-machine-body interface (BMBI) applications', Proceedings of the IEEE ISCAS, Vol. 1, pp. 1766–1769, 2015.

[7] F.J. Kao, G. Keiser, A. Gogoi, 'Advanced Optical Methods for Brain Imaging', Springer – Singapore, 2019, DOI 10.1007/978-981-10-9020-2.

[8] C. Hu, R. Sam, M. Shan, V. Nastasa, M. Wang, T. Kim, M. Gillette, P. Sengupta, G. Popescu, 'Optical excitation and detection of neuronal activity', Journal of Biophotonics, 12, e201800269, 2019.

[9] G.N. Angotzia, F. Boia, A. Lecomtea, E. Mielea, M. Malerbaa, S. Zuccab, A. Casilec, L. Berdondinia, 'SiNAPS: An implantable active pixel sensor CMOS-probe for simultaneous large-scale neural recordings', Biosensors and Bioelectronics, 126, pp. 355–364, 2019.

[10] G. Hong, X. Yang, T. Zhou, C.M. Lieber, 'Mesh electronics: A new paradigm for tissue-like brain', Current Opinion in Neurobiology, 50, pp. 33–41, 2018.

[11] N.A. Steinmetz, C. Koch, K.D. Harris, M. Carandini, 'Challenges and opportunities for large-scale electrophysiology with Neuropixels probes', Current Opinion in Neurobiology, 50, pp. 92–100, 2018.

[12] G. Hong, C. M. Lieber, 'Novel electrode technologies for neural recordings', National Reviews Neuroscience, 20, pp. 330–345, 2019.

[13] H.T. Lancashire, D. Jiang, A. Demosthenous, N. Donaldson, 'An ASIC for recording and stimulation in stacked microchannel neural interfaces', IEEE Transactions on Biomedical Circuits and Systems, 13, pp. 259–270, 2019.

[14] J. Xu, A. Tuan Nguyen, Z. Yang, 'Advances in neural recording and stimulation devices', in *Engineering in Medicine – Advances and Challenges*, Paul A. Iaizzo, Editor, Academic Press, pp. 335–363, 2019, DOI 10.1016/C2016-0-04120-7.

[15] J. Dong, B. Geng, I.K. Niazi, I. Amjad, S. Dosen, W. Jensen, E.N. Kamavuako, 'The variability of psychophysical parameters following surface and subdermal stimulation: A multiday study in amputees', IEEE Transactions on Neural Systems and Rehabilitation Engineering, 28(1), pp. 174–180, 2020.

[16] K. Li, P. Boyd, Y. Zhou, Z. Ju, H. Liu, 'Electrotactile feedback in a virtual hand rehabilitation platform: Evaluation and implementation', IEEE Transactions on Automation Science and Engineering, 16(4), pp. 1556–1565, 2019.

[17] A. Peer, G.A. Illan, H. Stüber, K.E. Friedl, I.R. Summers, 'A simulation environment for studying transcutaneous electrotactile stimulation', PLoS ONE, 14(2), e0212479, 2019.

[18] V. Amoli, J.S. Kim, E. Jee, Y.S. Chung, S.Y. Kim, J. Koo, H. Choi, Y. Kim, D.H. Kim, 'A bioinspired hydrogen bond-triggered ultrasensitive ionic mechanoreceptor skin', Nature Communications, 10(1), 4019, 2019.

[19] G. Liang, Z. Ruan, Z. Liu, H. Li, Z. Wang, Z. Tang, F. Mo, Q. Yang, L. Ma, D. Wang, C. Zhi, 'Toward multifunctional and wearable smart skins with energy-harvesting, touch-sensing, and exteroception-visualizing capabilities by an all-polymer design', Advanced Electronic Materials, 5(10), 1900553, 2019.

[20] A. Llamosi, S. Toussaint, 'Measuring force intensity and direction with a spatially resolved soft sensor for biomechanics and robotic haptic capability', Soft Robotics, 6(3), pp. 346–355, 2019.

[21] A. De Marcellis, E. Palange, M. Janneh, C. Rizza, A. Ciattoni, S. Mengali, 'Optimization of the detection sensitivity of plasmonic nanoantenna based sensors for mid-infrared spectroscopic applications', Proceedings of the EUROSENSORS 2015, Vol. 120, pp. 1179–1182, 6–9 September 2015, Freiburg (Germany), September 2015.

[22] N. Wang, J. Tong, W. Zhou, W. Jiang, J. Li, X. Dong, S. Hu, 'Novel quadruple-band microwave metamaterial absorber', IEEE Photonics Journal, 7, pp. 1–7, 2015.

[23] J.J. Jun, N.A. Steinmetz, J.H. Siegle, D.J. Denman, M. Bauza, B. Barbarits, et al., 'Fully integrated silicon probes for high-density recording of neural activity', Nature, 551, pp. 232–236, 2017.

[24] N.A Steinmetz, C. Koch, K.D. Harris, M. Carandini, 'Challenges and opportunities for large-scale electrophysiology with Neuropixels probes', Current Opinion in Neurobiology, 50, pp. 92–100, 2018.

[25] G. Hong, X. Yang, T. Zhou, C.M. Lieber, 'Mesh electronics: A new paradigm for tissue-like brain', Current Opinion in Neurobiology, 50, pp. 33–41, 2018.

[26] G.N. Angotzia, F. Boia, A. Lecomtea, E. Mielea, M. Malerbaa, S. Zuccab, et al., 'SiNAPS: An implantable active pixel sensor CMOS-probe for simultaneous large-scale neural recordings', Biosensors and Bioelectronics, 126, pp. 355–364, 2019.

[27] N.J. Cleven, J.A. Muntjes, H. Fassbender, U. Urban, M. Gortz, H. Vogt, et al., 'A novel fully implantable wireless sensor system for monitoring hypertension patients', IEEE Transactions on Biomedical Engineering 59, pp. 3124–3130, 2012.

[28] S. Park, D.A. Borton, M. Kang, A.V. Nurmikko, Y.-K. Song, 'An implantable neural sensing microsystem with fiber-optic data transmission and power delivery', Sensors, 13, pp. 6014–6031, 2013.

[29] A.D. Rush, P.R. Troyk, 'A power and data link for a wireless-implanted neural recording system', IEEE Transactions on Biomedical Engineering, 59, pp. 3255–3262, 2012.

[30] U.M. Jow, M. Ghovanloo, 'Optimization of data coils in a multiband wireless link for neuroprosthetic implantable devices', IEEE Transactions on Biomedical Circuits and Systems, 4, pp. 301–310, 2010.

[31] F. Zhang, M. Aghagolzadeh, K. Oweiss, 'A fully implantable, programmable and multimodal neuroprocessor for wireless, cortically controlled brain-machine interface applications', Journal of Signal Processing Systems, 69, pp. 351–361, 2012.

[32] R.K. Dokania, X.Y. Wang, S.G. Tallur, A. Apsel, 'A low power impulse radio design for body-area-networks', IEEE Transactions on Circuits and SysTems I, 58, pp. 1458–1469, 2011.

[33] Y. Gao, Y. Zheng, S. Diao, W. Toh, C. Ang, M. Je, et al., 'Low-power ultra-wideband wireless telemetry transceiver for medical

sensor application', IEEE Transactions on Biomedical EngIneering, 58, pp. 1291–1299, 2011.

[34] H. Miranda, V. Gilja, C.A. Chestek, K.V. Shenoy, T.H. Meng, 'HermesD: A high-rate long-range wireless transmission system for simultaneous multichannel neural recording applications', IEEE Transactions on Biomedical Circuits and Systems, 4, pp. 181–191, 2010.

[35] A. Ebrazeh, P. Mohseni, '30 pJ/b, 67 Mbps, centimeter-to-meter range data telemetry with an IR-UWB wireless link', IEEE Transactions on Biomedical Circuits and Systems, 9, pp. 362–369, 2015.

[36] M. Elzeftawi, L. Theogarajan, 'A 10pJ/bit 135 Mbps IR-UWB transmitter using pulse position modulation and with on-chip LDO regulator in 0.13 μm CMOS for biomedical implants', Proceedings of IEEE BioWireleSS, Vol. 1, pp. 37–39, 2013.

[37] K. Shimizu, 'Optical biotelemetry', in *Handbook of Biomedical Telemetry*, K.S. Nikita, Editor, Wiley-IEEE Press, Piscataway, New Jersey (USA), Chapter 10, pp. 301–329, 2014.

[38] T. Liu, U. Bihr, S.M. Anis, M. Ortmanns, 'Optical transcutaneous link for low power, high data rate telemetry', Proceedings of IEEE EMBS, Vol. 1, pp. 3535–3538, 2012.

[39] T. Liu, U. Bihr, J. Anders, M. Ortmanns, 'Performance evaluation of a low power optical wireless link for biomedical data transfer', Proceedings of IEEE BioCAS, Vol. 1, pp. 870–873, 2014.

[40] T. Liu, U. Bihr, J. Becker, J. Anders, M. Ortmanns, 'In vivo verification of a 100 Mbps transcutaneous optical telemetric link', Proceedings of IEEE BioCAS, Vol. 1, pp. 580–583, 2014.

[41] F. Inanlou, M. Ghovanloo, 'Wideband near-field data transmission using pulse harmonic modulation', IEEE Transactions on Circuits and Systems I, 58, pp. 186–195, 2011.

[42] H. Miranda, T.H. Meng, 'A programmable pulse UWB transmitter with 34% energy efficiency for multichannel neuro-recording systems', Proceedings of IEEE CICC, Vol. 1, pp. 1–4, 2010.

[43] A. De Marcellis, E. Palange, L. Nubile, M. Faccio, G. Di Patrizio Stanchieri, T.G. Constandinou, 'A pulsed coding technique based on optical UWB modulation for high data rate low power wireless implantable biotelemetry', MDPI Electronics, 5(4), pp. 69–78, 2016.

[44] A. De Marcellis, E. Palange, M. Faccio, G. Di Patrizio Stanchieri, T.G. Constandinou, 'A 250 Mbps 24 pJ/bit UWB-inspired optical communication system for bioimplants', Proceedings of the IEEE BioCAS 2017, Turin (Italy), Vol. 1, pp. 1–4, 19–21 October 2017.

[45] A. De Marcellis, M. Faccio, E. Palange, 'A 0.35 μm CMOS 200 kHz – 2 GHz Fully-analogue closed-loop circuit for continuous-time *Clock* duty-cycle correction in integrated digital systems', Proceedings of the IEEE ISCAS 2018, Vol. 1, pp. 1–5, Florence, May 2018.

[46] Y.-K. Lo, K. Chen, P. Gad, W. Liu, 'A fully-integrated high-compliance voltage SoC for epi-retinal and neural prostheses', IEEE Transactions on Biomedical Circuits and Systems, 7(6), pp. 761–772, December 2013.

[47] S. Hareendar, R. Jeya Raghul, A. Kumaravelan, 'Brain computer interface for controlling mobile robot', International Journal of Recent Technology and Engineering, 8(2), pp. 152–156, July 2019.

[48] I. Galkin, M. Vorobyov, O. Gainutdinovs, P. Studers, 'Customized vibration generator for state of health monitoring of prosthetic implants and pseudo-bionic machine-human feedbacks', MDPI Electronics, 8(7), 810, July 2019.

[49] S. Tam, G. Bilodeau, J. Brown, G. Gagnon-Turcotte, A. Campeau-Lecours, B. Gosselin, B., 'A Wearable Wireless Armband Sensor for High-Density Surface Electromyography Recording', IEEE Proceedings of the Annual International Conference of Engineering in Medicine and Biology Society (EMBS), pp. 6040–6044, 23–27 July 2019, Berlin (Germany).

[50] A.D. Rush, P.R. Troyk, 'A power and data link for a wireless-implanted neural recording system', IEEE Transactions on Biomedical Engineering, 59, pp. 3255–3262, 2012.

[51] U.M. Jow, M. Ghovanloo, 'Optimization of data coils in a multiband wireless link for neuroprosthetic implantable devices', IEEE Transactions on Biomedical Circuits and Systems, 4, pp. 301–310, 2010.

[52] F. Zhang, M. Aghagolzadeh, K. Oweiss, 'A fully implantable, programmable and multimodal neuroprocessor for wireless, cortically controlled brain-machine interface applications', Journal of Signal Processing Systems, 69, pp. 351–361, 2012.

[53] R.K. Dokania, X.Y. Wang, S.G. Tallur, A. Apsel, 'A low power impulse radio design for body-area-networks', IEEE Transactions on Circuits and Systems I, 58, pp. 1458–1469, 2011.

[54] A.M. Sodagar, K.D. Wise, K. Najafi, 'A wireless implantable microsystem for multichannel neural recording', IEEE Transactions on Microwave Theory and Techniques, 57, pp. 2565–2573, 2009.

[55] K. Duncan, R. Etienne-Cummings, 'Selecting a safe power level for an indoor implanted UWB wireless biotelemetry link', Proceedings of the

IEEE BioCAS 2013, Rotterdam, Netherlands, pp. 230–233, 31 Oct.–2 Nov. 2013.

[56] J. Kim, Y. Rahmat-Samii, 'Electromagnetic interactions between biological tissues and implantable biotelemetry systems', Proceedings of the Microwave Symposium Digest 2015, Long Beach, CA (USA), pp. 1–4, 17–17 June 2005.

[57] M. Haas, B. Schweizer, J. Anders, M. Ortmanns, 'A miniaturized UWB antenna for implantable data telemetry', Proceedings of the IEEE International Conference of Engineering in Medicine and Biology 2017, Seogwipo, South Korea, pp. 1086–1089, 11–15 July 2017.

[58] J. Kang, J. Lee, W. Lee, S.J. Kim, S.C. Kim, 'A miniaturized implantable dual-band platform design for wireless power transfer and biotelemetry', Proceedings of the Asia-Pacific Microwave Conference 2017, Kuala Lumpar, Malaysia, pp. 330–333, 13–16 Nov. 2017.

[59] Y. Zhang, C. Liu, X. Liu, K. Zhang, X. Yang, 'A wideband circularly polarized implantable antenna for 915 MHz ISM-band biotelemetry devices', IEEE Antennas and Wireless Propagation Letters, 17(8), pp. 1473–1477, 2018.

[60] A. De Marcellis, G. Di Patrizio Stanchieri, M. Faccio, E. Palange, P. Motto Ros, M. Martina, et al., 'Live demonstration: Event-driven serial communication on optical fiber', Proceedings of the IEEE ISCAS 2019, Vol. 1, pp. 1–5, Sapporo, May 2019.

[61] A. De Marcellis, G. Di Patrizio Stanchieri, P. Motto Ros, M. Martina, D. Demarchi, C. Bartolozzi, et al., 'Impulse-based asynchronous serial communication protocol on optical fiber link for AER systems', Proceedings of the IEEE International Conference on Electronics, Circuits and Systems (ICECS), Vol. 1, pp. 1–5, Genova, 27–29 November 2019.

[62] G. Di Patrizio Stanchieri, M. Saleh, M. Sciulli, A. De Marcellis, A. Ibrahim, M. Valle, et al., 'FPGA-based tactile sensory feedback system with optical fiber data communication link for prosthetic applications', Proceedings of the IEEE International Conference on Electronics, Circuits and Systems (ICECS), Vol. 1, pp. 1–5, Genova, 27–29 November 2019.

[63] M. Saleh, G. Di Patrizio Stanchieri, M. Sciulli, A. De Marcellis, Y. Abbass, A. Ibrahim, et al., 'Live demonstration: Tactile sensory feedback system based on UWB optical link for prosthetics', Proceedings of the IEEE BioCAS 2019, Vol. 1, pp. 1–5, Nara (Japan), 17–19 October 2019.

[64] E. Rodriguez-Villegas, 'A low-power wide-range I-V converter for amperometric sensing applications', IEEE Transactions on Biomedical Circuits and Systems, 3, pp. 432–436, 2009.

[65] E. Kamrani, F. Lesage, M. Sawan, 'A low-power photon-counter front-end dedicated to NIRS brain imaging', IEEE Sensors Journal, 15, pp. 3724–3733, 2015.

[66] T. Liu, Z. Cai, J. Anders, M. Ortmanns, 'A bootstrap transimpedance amplifier for high speed optical transcutaneous wireless links', Proceedings of the PRIME 2014, Grenoble, France, pp. 1–4, 30 June-3 July 2014.

[67] X. Li, P. Baltus, P. Van Zeij, D. Milosevic, A. Van Roermund, 'A 70 GHz 10.2 mW self-demodulator for OOK modulation in 65-nm CMOS technology', Proceedings of the IEEE CICC 2010, San Jose, CA (USA), pp. 1–4, 19–22 Sept. 2010.

[68] J. Jung, S. Zhu, P. Liu, Y.-J. Emery Chen, D. Heo, '22-pj/bit energy-efficient 2.4-GHz implantable OOK transmitter for wireless biotelemetry systems: In Vitro experiments using rat skin-mimic', IEEE Transactions on Microwave Theory and Techniques, 58, pp. 4102–4111, 2010.

[69] Y.-H. Wen, K.-M. Feng, 'A simple NRZ-OOK to PDM RZ-QPSK optical modulation format conversion by bidirectional XPM', IEEE Photonics Technology Letters, 27, pp. 935–938, 2015.

[70] K. Chi, Y.-H. Zhu, X. Jiang, V.C.M. Leung, 'Energy-efficient prefix-free codes for wireless nano-sensor networks using OOK modulation', IEEE Transactions on Wireless Communications, 13, pp. 2670–2682, 2014.

[71] M. Crepaldi, C. Li, J.R. Fernandes, P.R. Kinget, 'An ultra-wideband impulse-radio transceiver chipset using synchronized-OOK modulation', IEEE Journal of Solid-State Circuits, 46, pp. 2284–2299, 2011.

[72] M.A. Hannan, S.M. Abbas, S.A. Samad, A. Hussain, 'Modulation techniques for biomedical implanted devices and their challenges', Sensors, 12, pp. 297–319, 2012.

[73] A. De Marcellis, E. Palange, V. Liberatore, L. Nubile, M. Faccio, T.G. Constandinou, 'A new modulation technique for high data rate low power UWB wireless optical communication in implantable biotelemetry systems', Proceedings of the EUROSENSORS 2015, Vol. 1, pp. 1–4, Freiburg (Germany), September 2015.

[74] A. Malekmohammadi, M.A. Elsherif, 'A novel multilevel coding technique for high speed optical fiber communication systems', Optik, 125, pp. 639–643, 2014.

[75] A. Mahdi, V. Paliouras, 'A low complexity-high throughput QC-LDPC encoder', IEEE Transactions on Signal Processing, 62, pp. 2696–2708, 2014.

[76] S. Saponara, 'Recent advances in integrated photonic-electronic technologies for high-speed processing and communication circuits for light-based transducers', Proceedings of the SPIE Defense + Commercial Sensing, Vol. 10996, Baltimore, 2019.

[77] S. Lourdudoss, J.E. Bowers, C. Jagadish, Editors, 'Future Directions in Silicon Photonics', Semiconductors and Semimetals, Vol. 101, 2019.

[78] A. De Marcellis, G. Di Patrizio Stanchieri, M. Faccio, E. Palange, T.G. Constandinou, 'A 300 Mbps 37 pJ/bit Pulsed Optical Biotelemetry', IEEE Transactions on Biomedical Circuits and Systems, Vol. 14, Issue 3, pp. 441–451, June 2020.

[79] A. De Marcellis, G. Di Patrizio Stanchieri, E. Palange, M. Faccio, T. G. Constandinou, 'An ultra-wideband-inspired system-on-chip for an optical bidirectional transcutaneous biotelemetry', Proceedings of the IEEE BioCAS, 2018, pp. 1–4, Cleveland (OHIO, USA), 17–21 October 2018.

[80] A.N. Bashkatov, E.A. Genina, V.I. Kochubey, V.V. Tuchin, 'Optical properties of human skin, subcutaneous and mucous tissues in the wavelength range from 400 to 2000 nm', Journal of Physics D: Applied Physics, 38, pp. 2543–2555, 2005.

[81] T. Lister, P.A. Wright, P.H. Chappell, 'Optical properties of human skin', Journal of Biomedical Optics 17, pp. 090901-1–090901-15, 2012.

[82] A. De Marcellis, G. Di Patrizio Stanchieri, E. Palange, M. Faccio, T.G. Constandinou, 'A 0.35 μm CMOS UWB-inspired bidirectional communication system-on-chip for transcutaneous optical biotelemetry links', Proceedings of the IEEE BioCAS 2019, Vol. 1, pp. 1–5, Nara (Japan), 17–19 October 2019.

9

Artificial Skin and Electrotactile Stimulation for Advanced Tactile Feedback in Myoelectric Prostheses

Lucia Seminara[1], Matija Strbac[2], Youssef Amin[1], Maurizio Valle[1] and Strahinja Dosen[3,*]

[1]CosmicLab, Department of Electrical, Electronic, Telecommunications Engineering and Naval Architecture (DITEN), University of Genoa, Italy
[2]Tecnalia Serbia Ltd., Belgrade, Serbia
[3]Department of Health Science and Technology, The Faculty of Medicine, Aalborg University, Denmark
E-mail: lucia.seminara@unige.it; matija.strbac@tecnalia.com; youssef.amin@edu.unige.it; maurizio.valle@unige.it; sdosen@hst.aau.dk
*Corresponding Author

Modern prosthetic arms and hands are sophisticated robotic devices that can provide some of the motor functions lost due to an amputation. However, none of the commonly used commercial systems restores somatosensory feedback to its user. In principle, the latter can be achieved by recording data from prosthesis sensors and conveying this information by stimulating the sensory structures of the amputee using invasive and noninvasive interfaces. Many such systems have been presented in the literature with promising results; however, they all rely on using only a few discrete stimulation points to transmit information and hence suffer from a limited information bandwidth. In this chapter, we propose a novel concept of a high-bandwidth feedback interface that relies on advanced sensing and stimulation to convey a large amount of information to the prosthesis user. The interface comprises an artificial skin covering the prosthesis with a dense network of tactile sensors (taxels) and a compact stimulation device delivering electrical current pulses through a matrix electrode with many conductive pads. The state-of-the-art in the two technologies that are required for the implementation of the proposed concept are reviewed. This includes biomimetic e-skins that are suitable for the application in a wearable scenario, stimulation systems integrating a demultiplexing circuit to distribute electrical pulses, and flexible electrodes that can be produced with arbitrary shape, size, and distribution of conductive pads. Finally, the challenges in selecting feedback variables (raw signals versus high-level features) and mapping of these variables into stimulation parameters are addressed.

9.1 Introduction

The loss of an upper limb is a dramatic event that profoundly changes the quality of life of the affected person. Human hands are sophisticated end effectors that enable dexterous grasping and manipulation and haptic exploration of the environment, and in addition, they are also an essential part of social and affective communication [1]. The ease with which humans control their hands to accomplish a variety of tasks has always fascinated the scientist working in the field of motor control [2]. The hands are characterized by a complex skeletal structure comprising 27 degrees of freedom (DoFs) actuated by a number of intrinsic and extrinsic muscles. The skin is endowed with a dense network of mechanoreceptors that react to different aspects of tactile stimuli, transmitting spatially distributed feedback to the central nervous system. The connection between the brain and the hand is mediated via an impressive network of motor and sensory nerves. In a recent anatomical study [3] that analyzed nerve specimen from nine human subjects, it was reported that approximately 350,000 nerve axons emerge from the spinal cord to innervate the human upper limb from the spinal cord, establishing sensory-motor communication.

The lost limb can be substituted by a powered prosthesis. By definition, the prosthesis is a morphological and functional substitute of a human limb. The former means that ideally the bionic limb should have similar size, shape, color, and weight as its biological counterpart, while the latter indicates that the artificial system needs to substitute for the lost functions. This includes not only motor functions, the ability to reach, grasp, and manipulate, but also somatosensory feedback that reflects mechanical interaction between the limb and environment. And indeed, it is well established in the literature on human motor control that feedback is indispensable for movement learning, adaptation, and execution [4]. The importance of feedback is strikingly expressed by the observation reported in the aforementioned anatomical study [3], namely, that sensory axons outnumber motor axons by a ratio of at least 9:1 in all the investigated nerves.

The lost motor functions can be restored by using myoelectric prostheses. These systems are controlled by recording the electrical activity of the user muscles (electromyography) to estimate his/her motor intention, which is then translated into commands for the prosthesis. The mechatronic and control methods and technologies are developing rapidly, especially in recent years. Nowadays, the prosthetic hands and arms approach the dexterity of biological limbs [5, 6] and integrate 3 DoF shoulder joints as well as the hands with individually controllable fingers [7]. The classic approach to prosthesis control in which an antagonistic muscle pair is used to move a single DoF and a switching signal (e.g., coactivation) to change across the DoFs is still the most prevalent commercially. Nevertheless, recently, the two companies detailed in [8] and [9] started offering the solutions based on pattern classification. In this approach, machine learning is employed to estimate the intended multi-DoF movement (e.g., grasp type) from multichannel electromyography (EMG) directly, eliminating thereby the need for the explicit switching across the DoFs.

The developments in mechatronics and control are in sharp contrast to the state-of-the-art in the restoration of somatosensory feedback. None of the commonly used prosthetic devices provides any tactile feedback to the user. In the reviews of user needs [10], however, the amputee subjects indicate that the restoration of feedback is indeed one of the important goals to improve the functionality as well as user experience. It is therefore not surprising that the restoration of feedback receives an increasing attention of the academic community as well as industry. Several recent reviews thoroughly document the impressive research efforts and developments in this field [11–13]. Two commercial systems that integrate a feedback interface have been presented recently [14, 15], but they are not widely used and their clinical efficiency is yet to be demonstrated.

The general approach to restoring the somatosensory feedback is to read sensor data from prosthesis and then convey this information to the user by stimulating his/her sensory structures, which are still available after the amputation. The stimulation can be delivered using invasive and noninvasive approaches. In the former case, the electrical pulses are delivered to the peripheral nerves to activate afferent fibers using electrodes that encircle [16] or penetrate the nerve [17, 18]. Since the stimulation activates the same nerves that previously innervated the lost hand, the elicited tactile sensations are somatotopic (i.e., felt as if they emanate from the "phantom" hand). To deliver the feedback noninvasively, the skin of the residual limb is stimulated mechanically or electrically. Mechanical stimulation can be delivered in the same modality as the feedback variable. For instance, prosthesis grasping force can be conveyed by pushing into [19] or squeezing [20] the residual limb. Alternatively, the feedback information can be transmitted using sensory substitution, as when grasping force is communicated by delivering vibrations using miniature vibromotors [21]. Finally, in electrical stimulation, low intensity electrical pulses are delivered to the skin to activate cutaneous afferents and elicit tactile sensations (electrotactile or electrocutaneous stimulation) [22].

Many feedback interfaces utilizing these approaches have been developed and tested [23], some even after a long-term use in the home environment [24], and the results were promising although multifaceted. More specifically, endowing the prosthesis with feedback is not guaranteed to improve the utility and performance. The overall benefit of feedback depends on multiple factors, including the characteristics of the control interface, system dynamics, complexity and nature of the task as well as the experience of the user [25]. Nevertheless, if implemented by considering the totality of the motor control loop, the feedback can indeed lead to significantly better performance as well as improved user experience [26].

9.2 High-Density Sensing and Stimulation

One point that is common to the solutions presented so far is that the tactile stimulation is delivered using only a few discrete stimulation channels. Typically, the feedback interface includes several vibration motors or stimulation electrodes, and this limits the quantity of information that can be transmitted to the user. Therefore, in most studies, the feedback is reduced to conveying a single feedback variable, usually the magnitude of the grasping force [9, 24, 25].

Importantly, such feedback is very different from the natural sensations coming from the biological limbs. For instance, when grasping an object, an intact-limb subject feels a pressure distribution characterizing the contact between the hand and the object. This is possible due to the advanced sensing and innervation of the biological hand. Different types of mechanoreceptors are densely distributed within the skin of the hand and the information from these sensors is transmitted to the central nervous system through an intricate network of afferent fibers.

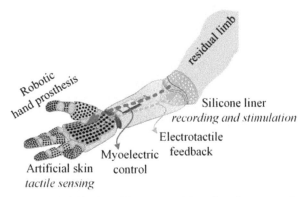

Figure 9.1 A high-bandwidth closed-loop control interface for a myoelectric prosthetic hand. The hand is covered by an artificial skin with a dense network of sensing points (black dots) while a flexible electrode comprising a matrix of conductive pads (white dots) is placed around the residual limb. The sensor data recorded by the skin are mapped into tactile sensations by delivering electrocutaneous stimulation through the electrode. The conductive pads are also used to record myoelectric signals for prosthesis control, hence a closed-loop interface.

Ideally, prostheses should be endowed with an artificial interface that mimics the characteristics of the natural feedback. This can be achieved by integrating advanced sensing and stimulation technology into the prosthetic system, as conceptualized in Figure 9.1. More specifically, we propose to endow the prosthetic hand with an artificial skin integrating a dense network of tactile sensors. This will allow measuring the mechanical interaction between the hand and the environment with high fidelity. At the same time, the residual limb of a prosthesis user will be covered with a flexible electrode integrating a high-density network of conductive pads. Such an interface will enable delivering versatile patterns of electrotactile stimulation that can be modulated in location (active pads) and parameters (intensity and frequency). The feedback will therefore elicit spatially distributed tactile sensations that can mimic the "feeling" produced by the natural feedback. For example, when the user commands the prosthesis to grasp an object, the prosthesis will close and the contact will be established in many points between the hand and the object, where the exact distribution of points depends on the characteristics of the hand (e.g., compliance, dexterity, etc.). The artificial skin will measure the properties of the contact (e.g., location and intensity) and this information will be transmitted to the user by delivering electrotactile stimulation through the matrix electrode. The user will feel tactile sensations spreading around his/her

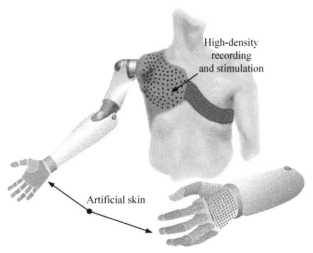

Figure 9.2   A high-bandwidth closed-loop interface for a full prosthetic arm. The hand is endowed by an e-skin as in Figure 9.1 and the stimulation is delivered to the skin of the chest area.

residual limb to mimic the distribution of the detected contacts as close as possible. The encoding algorithm programmed in the embedded stimulator will determine the stimulation parameters (intensity and active pads) that need to be used to produce such sensation.

The proposed sensing and stimulation interface will therefore provide a high-bandwidth connection between the brain and the bionic limb, far beyond what is possible in the state-of-the-art. Importantly, this can be achieved using noninvasive technology, which is very relevant considering that some amputees might be reluctant to undergo additional surgery. The same electrode matrix could be employed to record EMG signals that are used for prosthesis control, which would also benefit from a dense network of signal detection points providing high-fidelity map of muscle activity.

A similar approach can be implemented also in case of a more complex system for a higher level of amputation, as illustrated in Figure 9.2. In this case, the stimulation has to be delivered to the upper arm (transhumeral amputation) or to the chest area (shoulder disarticulation). In both figures, the e-skin is placed at the hand, since this is most critical for grasping and manipulation, however, it can be easily envisioned that the artificial skin could cover other parts of the prosthetic system, ideally, the whole prosthesis (in the same way the natural skin covers the arm).

In the following sections of this chapter, we present the development of the components that are required to design the proposed high-bandwidth closed-loop prosthetic system, namely, electronic skins that are suitable for wearable applications (Section 9.3) and the multichannel stimulation systems that can be used to deliver electrocutaneous stimulation through matrix electrodes (Section 9.4).

9.3 Electronic Skin Systems for Prosthetics

Touch – or somatosensation – is at the core of any mechanical interaction. The main challenge to artificially reconstruct the sense of touch relates to shaping brain-controlled prosthetic devices to be capable to handle real-world objects in open-ended conditions and communicate effectively with the human nervous system through a stable and robust bidirectional interface. A recent special issue [27] provided a wide overall view on state-of-the-art theoretical and applied studies related to the development of flexible electronic skins. Chapter 2 (Artificial Tactile Sensing and Electronic Skin Technologies), also, deals with artificial tactile sensing and electronic skin technologies. The reader is referred to that chapter for the description of human skin physiology and history of artificial skins. The current text will specifically highlight those aspects of the human touch that have been artificially mimicked and it will recall the main challenges beyond sensor design, as these are key points for the development of human-in-the-loop prosthetic devices sensitive to touch. We will then focus on a number of distributed sensing systems that have been particularly conceived or are promising for the application in prosthetics.

9.3.1 Biomimetic e-skins for Prosthetic Systems

The core of human perceptual abilities related to touch is first connected to the properties of single mechanoreceptor types, in that they constitute the mechanical sensory interface between the human body and the surrounding environment [28]. Nerve fibers that end in Meissner and Pacinian corpuscles are fast adapting (FA, responding to vibrations and force derivative), while those that end in Merkel cell complexes and Ruffini endings are slow-adapting (SA, producing a sustained signal in response to a sustained stimulus) [29]. Likewise, an e-skin should be provided with tactile sensors to detect both static (e.g., using capacitive, resistive, optical, magnetic, and inductive technology) and dynamic (e.g., using piezoelectric or triboelectric principle) contact events. Several reviews compared touch sensors based on their transduction mechanisms and a broad overview of materials and sensing technologies for advanced biomimetic tactile sensing for prosthetics is reported in [30]. Most common transducers are capacitive or piezoresistive to measure skin deformation and piezoelectric for pressure and vibration measurement. The former mainly mimic SA receptors in humans [30]. More specifically, Merkel cell neurite complexes (SA I) are located near the skin surface and respond to skin indentations with high sensitivity. They are densely distributed in sensitive areas of the skin, such as the fingertips, and they consequently provide high-resolution force information, which is useful to respond to fine spatial details (e.g., edges) and for texture and pattern perception. On the other side, Ruffini endings (SA II) respond to sustained pressure (0.4–3 Hz) and are primarily responsible for measuring spatial deformation (particularly lateral skin stretch). Capacitive sensors have been used to detect static/quasi-static events with high sensitivity, low power consumption, simple device architecture, and simple readout electronics [31]. The large-area e-skin to cover the surface of the well-known iCub robot is based on this transduction mechanism [32]. On the other hand, piezoelectric sensors have been employed to detect dynamic events [33] and are attractive as they could lead to self-powered sensors [34]. Due to their large bandwidth, they potentially mimic all mechanoreceptors in the skin and are in principle appropriate to handle a large variety of functions. Together with SA receptors described earlier, they also mimic Meissner corpuscles (FA I) measuring skin strain rate and responding to low-frequency vibrations (3–40 Hz), such as those induced in the skin by object slipping across the fingers and typically associated with object manipulation, grip control, and texture discrimination [30]. Finally, the bandwidth of a piezoelectric sensor is also compatible with Pacinian corpuscles (FA II) detecting high-frequency vibrations (40–500 Hz), such as those induced upon insect landing on the skin or by tapping a rigid tool on the table, and are again important for texture discrimination and slip detection [30]. To make a practical example, the light indentation of the glabrous skin with a pointed object elicits a response from FA receptors near the contact. This response is most of the time transitory, i.e., stronger during the ramping periods of the indentation at the beginning and the end of the stimulus. Similarly, artificial piezoelectric sensors respond with bursts of potentials when the stimulus is applied and removed.

Tactile information processed in the somatosensory areas supports both perception and action, as it both transmits perceptual information on the touched object and provides feedback to the motor system. Together with mechanoreceptors, the temperature sensing and pain can be also artificially mimicked. The former is commonly realized by thermistors or temperature-sensitive resistors. Nociception transduction of a painful stimulus has been only implemented by detecting high-intensity signals through tactile (during interaction with a sharp object) and temperature (while touching extremely hot or cold objects) sensors [35]. Proprioception, referring to the sense of self-movement and body position, is also relevant when dealing with active tasks and closing the sensorimotor control loop in prosthetics. Recent studies [36] hint at touch providing auxiliary proprioceptive feedback for guiding actions in humans, which might inspire novel ways for haptic feedback in prosthetics. More traditional approaches for proprioception adopt discrete sensors most commonly based on inertial measurement units (IMUs), measuring linear acceleration, angular velocity and orientation, stretch sensors for joint movements, and magnetic sensing to determine orientation [35]. All these sensors can be potentially embedded into an e-skin to transduce various modalities of environmental stimuli with the ambitious goal of restoring the dexterity and natural sensations of a biological hand. The reader is referred to [35] for a summary table of main sensor types, transduction methods, and a number of corresponding references interesting for the prospective application of e-skins in prosthetics.

Lederman and Klatzky [37] pointed out that the foundation of human abilities in processing touch information resides in the sensory primitives that are signaled by the peripheral receptors. This is also the reason why mechanoreceptors have inspired artificial transduction mechanisms. However, it is worth noting that no complete models are available yet to predict behaviors from tactile functions and responses from single receptors. Interesting open issues still relate to how the mechanical state of the human skin defines the relationships between skin mechanoreceptors and how the resulting spatiotemporal patterns of receptor activation are related to the brain cortical activity [31]. It is posited that sensory characteristics are not a property of the specific mechanoreceptor, but of the complete mechanical structure and the way in which a specific mechanoreceptor is integrated into it. For example, clusters of slowly adapting (SAII) mechanoreceptors in the skin folds bordering the nail and corresponding afferents encode contact force vectors [38], while analogous mechanoreceptors embedded in the soft skin layer measure skin stretch. Analogously, coupling piezoelectric polymer PVDF sensors to a rigid substrate makes their response directly proportional to the normal stress component (i.e., pressure) [33], while interpreting their response when they are coupled to a deformable substrate requires a more complicated model as sensors are subjected to bending [39]. As for the substrate, the same reasoning can be applied to the outer skin layers: interaction through contact involves traction distributions at the surface of the skin that have effects which are more dramatic when the skin is compliant. Due to its softness and compliance, the human skin interface conforms to the surface of the touched object, altering the force distribution transmitted to embedded mechanoreceptors [40], increasing contact area, and improving grasping through friction [41]. The surface properties of the skin are also to be taken into account, for instance, fingerprint ridges are instrumental for texture discrimination [42]. In addition, the invasive quality of the mechanical interaction increases the complexity of the design of artificial sensing systems, requiring robustness and resilience to wear and abrasion. All in all, the skin structure, its morphology, micromechanical, and tribology properties modulate the collective response of a large number of receptors during the interaction of the human skin with the environment [31].

This unique combination of mechanical and sensory properties did not exist in conventional electronics and the insights from the biology paved the way for the development of new biomimetic materials and processing methods [43], leading to deformable sensors and flexible/stretchable electronics [44–47]. There are comprehensive developments in this field, including manufacturing technologies, deployment processes and processing methods to fabricate sensors and electronics over flexible substrates [48, 49], stretchable electronics to mimic and adapt to the skin [47], materials with low elastic moduli and good stretchability to build skin-like coverings for prosthetic devices with similar feel (modulus, viscoelasticity, and temperature) as real skin [50], and novel composite materials to mimic the skin toughness and durability [51], to cite some examples. For a thorough overview of these aspects the reader is referred to [30, 31], and references therein. It is worth noting that measuring the mechanical state of the skin (three-dimensional [3D] continuous medium) by a finite number of sensors generating scalar outputs causes substantial dimensionality reduction. Flexible and stretchable technologies could be also used to embed sensors with different properties in an elastic material at different depths and better map the mechanical state of the e-skin, though this approach has been rarely pursued [31].

Observing from the perspective of the perceptual qualities of the whole skin structure, spatial- and temporal-resolving capacities of the skin are relevant for the haptic perception of object and surface properties. The fact that single mechanoreceptors are organized into dense networks provides spatially distributed information in humans. Higher mechanoreceptor and innervation densities in sensitive areas of the skin, such as the fingertips, provide high-resolution force information to respond to fine spatial details with high spatial acuity [52]. The resolution, density, and response time of human skin have been mimicked with different sensor architectures and readout interfaces [31]. Both organic and inorganic semiconductor-based sensors have been explored to realize high-spatial resolution field-effect transistor (FET)-based sensor arrays on planar or flexible substrates [44, 53]. Adequate spatial resolution is necessary for perception of surface roughness and textures at different scales [54, 55]: a biomimetic approach to achieve an unprecedented level of performance in identifying objects by their tactile texture has been developed using a biologically inspired tactile sensing system (BioTac) [56]. Vibrations of the skin are induced when sliding over textured surfaces and they propagate efficiently through the liquid-filled sensor where they can be measured by a pressure sensor. Fingerprint-like ridges with biomimetic size (0.4 mm spacing) enhance detected vibrations [57] and are expected to also enhance local tactile shape discrimination [58], in line with previous consideration that surface properties of the structural skin layers affect the sensing properties of the human touch. The spatiotemporal resolving power of the same finger sensing system has also been tested in measuring its ability of localizing contact and discriminating pressure, motion direction, and speed (together with textures) [59].

As just shown with the example of the BioTac finger, to endow haptic devices with human-like perception and action abilities to manipulate objects, they must be provided not only with tactile sensors but also with intelligent algorithms to select, control, and interpret sensor data. This may require solving such open issues as how to handle high sensor data rates (many sensors, large bandwidth), how to make sensor fusion, and how to extract information intelligently out of the multitude of sensor data. All this affects timing for the sensory feedback to the nervous system and may strongly influence the embodiment of a prosthesis: in humans, the time delay for tactile signals to reach the brain is in the range of tens of milliseconds [60]. Some interesting elements might be adopted from robotics, which addresses such issues as the representation and management of heterogeneous, distributed sensory information and the adoption of a real-time computational infrastructure to collect, process, and transfer sensory data [61]. To avoid high communication bandwidth, the e-skin can process raw tactile data close to the sensor location and extract structured information by machine learning paradigms. High-level features extracted from tactile sensors for sensorimotor control are mainly normal forces useful for grasp control, object manipulation, and orientation determination; tensile strains for proprioception; shear forces for grasp control and friction determination; and vibrations for slip detection and texture determination [61]. However, embedding machine learning imposes challenges in terms of time latency, energy consumption, and memory storage [62]. This issue can be addressed by relying on approximate computing techniques applied at algorithmic, architecture, and circuit levels [63]. Decoupling approximate computing methods with hardware accelerators used in traditional processors may also improve the efficiency of such systems. Moreover, taking benefit of the parallelism featured in parallel ultralow power (PULP) processors is one of the most employed solutions to tackle these challenges [64].

In humans, the electrical pulses (action potentials) encode information on the mechanical interaction and allow for long-distance information transfer. Time-continuous voltage signals can be transmitted in artificial sensing systems, though conventional tactile systems transmitting "always-on" information serially lead to issues discussed earlier. Alternative more biomimetic approaches for human-in-the-loop sensorimotor systems rely on neuromorphic tactile skins, which are expected to bring some fresh contributions to this field. Biological neural sensory and processing systems rely on the *neuron* as the processing element, which accumulates input spikes for a certain time, produces its output spike, and resets itself. Synapses connect different neurons and transport spikes, introducing a weight: they thus perform communication and memory functions simultaneously. Processing is performed together by neurons (typically grouped into populations) and synapses, preventing a clear distinction between elements performing processing, communication, and memory tasks. Neuromorphic systems try to mimic the efficient and low-power sensing and information processing of the human nervous system, though the performance of these artificial systems is still well below that of their biological counterparts [65]. In a neuromorphic e-skin, the tactile signal is converted into event-driven sensing strategies for efficient encoding, transmission, and processing of tactile information. The event-driven paradigm should be applied at the level of sensory acquisition and consists of a first type of tactile sensors (taxels) detecting *transient* information (onset of contact and contact release) and another type of taxels encoding the pressure level through their instantaneous firing rate [66]. A very recent work merges different perspectives on event-driven systems and evaluates the applicability of existing event-driven implementations in large-area e-skin systems [67]. To give an example relevant to prosthetics, in [68] the frequency of digital output signals encodes the intensity of the mechanical interaction in organic semiconductor-based sensing devices mimicking slow-adapting mechanoreceptors. The reader is referred to [4] for aspects related to the electronics for reading sensory data and encoding biomimetic output. Specific neuromimetic architectures can simultaneously transmit tactile information while preserving low latencies, thereby resolving fine-grained spatiotemporal features necessary for rapid tactile perception [69]. Hardware-based neuromorphic tactile data processing with neural network-like algorithms would be more energy efficient than software-programmed neural networks [4]. The reader is referred to [4] and references therein for examples of devices imitating biological neuron functioning and acting as possible building blocks for such hardware-based neural networks, e.g., the neural nanowire FET device, or neuromorphic architectures such as memristors, spin logic, neuron MOSFET or analogue circuit-based neurons, though they have not been adequately explored yet for tactile skins.

The last point concerns power management and energy autonomy. Even if the surface of the prosthetic device is not as big as the whole surface of the robot body, covering a prosthetic limb with high density of multiple types of electronic components (e.g., sensors, actuators, electronics, etc.) requires energy. The need for frequently recharging batteries or energy storage devices is certainly an obstacle for amputees to accept integrating advanced prosthetic devices into their everyday life. Therefore, flexible energy systems to realize energy-autonomous e-skin are relevant solutions for prosthetics applications. A few preliminary designs have been proposed, consisting of sensing systems including energy generators such as piezoelectric (mechanical), solar microcell arrays (light), biofuel cells (chemical), storage devices (stretchable batteries and self-capacitors), and high efficiency power transmission systems [70]. On the sensing side, graphene-based sensors look promising [71]. There have been already different attempts to harvest ambient energy for self-powered e-skin [72]. Obviously, energy management strategies are needed to reduce power demand by, for example, dynamically addressing only those sensors relevant for the task at hand. Alternatively, some recent attempts for power-optimized system implementations targeted wireless power transfer and battery-less operations. Details on all these aspects are illustrated in [31, 70], and references therein. To account for this critical issue in e-skin development, we include a reference example of self-powered e-skin in Table 9.1.

Table 9.1   Representative electronic skin systems targeting prosthetic applications

References	Sensing	Processing and Encoding	Suitable Features
[74]	**Supramolecular biomimetic e-skin** • Transparent hydrogel skin • Capacitive and resistive sensors	**Output** • Finger bending–straightening from capacitance changes upon deformation • Compression stress from geometry changes • Temperature increase from real-time resistance decrease	**Sensing** • *Multimodal*: Temperature, strain, and stress sensing capabilities • high sensitivity • wide temperature range **System** • compliant/ flexible/ stretchable • self-healing • robust
[75]	**Stretchable and conformable matrix networks** (SCMN) • 5 SCMN with pressure sensors (1 middle, 3 forefinger, 1 thumb) • 1 SCMN with temperature sensor (thumb) • 6 sensor types fabricated and tested	**Output** • Relative humidity • UV photodetection • Static/dynamic magnetic field • Pressure/grip strength and proximity • Temperature, bending and in-plane strain	**Sensing** (multimodal arrays) • Detect and differentiate three or more stimuli simultaneously • Adjustable sensing range • Large-area expandability • Highly stretchable • Detect pressure as low as 7.3 ± 1.2 Pa • Proximity sensing from 4 cm • High-density 3D integration **System:** coupled to a robotic hand

(Continued)

Table 9.1 Continued

References	Sensing	Processing and Encoding	Suitable Features
[76]	**Stretchable optical waveguides sensor** • 3 waveguides into each finger actuator-final waveguide acts as touch sensor for the fingertip • LEDs transmit light and photodetectors can sense it: light power loss is related to strain • Each digit is capable of both proprioception and exteroception	**Output** • Bending (+ direction) • Elongation • Curvature • Pressure Signal processing to detect: • shape • texture • softness • object (classification)	**Sensing** (proprioceptive and tactile) • Highly repeatable output • Easy to calibrate • Chemically inert • Low hysteresis • High precision • High sensitivity • Compatible with large strains • Complex sensor shapes possible **System:** coupled to a prosthetic hand • Easy to fabricate • Highly compliant and stretchable (~85% strain • High SNR >50
[5]	**Modular prosthetic Limb** • Fingertip sensors to sense force (strain gauge/capacitive array), vibration (3-axis accelerometers), heat flux (thermistor), contact (PVDF piezoelectric polymer array) • Join proprioception	**Output** • Position, velocity, torques, and internal temperature feedback for each actuated joint • From contact sensors: localization of applied forces and feature discrimination • From accelerometers: vibration, surface texture recognition	**Sensing** (proprioceptive and tactile) • *Multimodal*: Temperature, strain, and stress sensing capabilities + proprioception • PVDF sensors: high sensitivity, mechanically durable, wide bandwidth, self-powered **System:** coupled to a prosthetic limb

9.3.2 Sense of Touch in Prosthetics: Case Studies

Toward the development of distributed sensing systems that can be interfaced intuitively with the human being, the PH-coding project (EU FET Open) [73] has been recently approved among those projects considered as the most challenging and breakthrough in the framework of current research. This project might cross the boundaries of what is currently possible, opening novel ways for haptic perception and natural movements with new-generation prosthetic devices, suitable to cope with incredibly rich and diversified representations in the brain. At the present stage, though, this goal seems out of reach, as many open issues have still to be managed for successfully coupling an artificial prosthetic skin to a nervous system. The components of the e-skin that need to be developed include sensors, signal encoding, signal transmission, and a method to convey the sensory information to the nervous system (feedback encoding) [30]. In Table 9.1, we summarize several examples of e-skin systems specifically conceived for the application in prosthetics by addressing issues such as sensors, electronics, manufacturing, and resilience.

A recent example of highly-sensitive multimodal skin, integrating distributed and heterogeneous sensors was presented in [74]. In particular, when this biomimetic skin is stimulated by both strain and temperature in a complex environment, the capacitive sensor detects the strain stimulus, while the temperature variation can be derived from the resistive sensor (with wide measurable range). This transparent hydrogel skin shows shape reconfiguration ability to adapt to irregular surfaces and is also compliant to prosthetic finger locomotion (stretchability). An important property of the system, substantially increasing the resilience, is that the skin mimics self-healability of natural skins: it not only shows elastic resilience under finger presses, but is also able to reconfigure its shapes and heal cracks very fast. However, the system has been implemented on a custom-made 3D printed prosthetic finger and not yet on a real prosthesis. Therefore, signal encoding and transmission together with a feedback method to send usable information to the user are still missing. Moreover, long-term stability is to be evaluated.

The next example, presented in [75], shows the potentialities of advanced sensing systems coupling stretchable technologies with a rich sensorization. Highly stretchable and conformable matrix networks (SCMN) are made of 100 nodes (1.6 mm in diameter), built on polyimide wires coated with Au thin film (50-μm width, 25-μm thickness). An intelligent prosthetic hand is equipped with personalized SCMN configurations on three fingers (thumb, forefinger, and middle finger), at present only including pressure and temperature sensors. Grip strength and temperature are retrieved from the sensors on the prosthetic hand, but closed-loop control, utilizing feedback, was not implemented. Overall, the feasibility of detecting temperature, in-plane strain, relative humidity (RH), ultraviolet (UV) light, magnetic field, pressure, and bending was demonstrated. The main drawback is that this system is neither robust to sensing conditions nor resilient to wear and abrasion.

Stretchable optical waveguides have been used for strain sensing in a soft and dexterous prosthetic hand [76]. The prosthetic hand is composed of four pneumatically actuated soft fingers and a thumb mounted onto a rigid palm. Each digit is a hollow silicone tube with fibers patterned into an elastomer. Compressed air enters each finger hollow core and causes the finger to bend and the hand to grasp. Photonic strain sensors are embedded into the actuator membranes and used as curvature, elongation, and force sensors, providing both proprioception (sensing internal pressure and active bending) and exteroception (sensing passive bending and external force at the fingertip) to each digit. The sensorized prosthetic hand has been employed to conduct various active sensation experiments inspired by the capabilities of a real hand, including detecting shape and texture, probing softness, and recognizing objects. All these experiments are purposely performed in open loop, to focus on the importance of the stretchable waveguides for active sensation. Future challenges relate to increasing the sensor number for both proprioception and exteroception and improving the sensor sensitivity. The system could be interesting for its sensory capabilities; however, no human-in-the-loop integration was demonstrated in the study.

The modular prosthetic limb (MPL) and its rich sensorization are thor-oughly illustrated in [5]. Each actuated joint in the MPL has position, velocity, torque, and internal temperature feedback. Through accelerometers and force sensors on fingertips, autonomous grasping (slip detection algorithm) and surface texture characterization could be directly implemented at the pros-thesis level. Low frequency force feedback provided by the fingertip sensors could be used for "human-in-the-loop" force control, though this has not been done yet.

BioTac is to be finally mentioned among most relevant biomimetic tactile sensing systems enabling human-like haptic perception suitable for prosthet-ics. It measures changes of impedance on its electrode array in response to force on the fingertip and integrates a thermistor to measure temperature. Tactile data processing enables fine discrimination of textures [56], detect-ing object compliance [77], and grasping fragile objects [78], to cite some representative examples. The reader is referred to the previous section for a more thorough description of the system.

In Table 9.2, we present two examples of a complete "e-skin sensing" – stimulation system to close the sensorimotor control loop in prosthetics. Together with issues related to sensing, Table 9.2 also summarizes the information related to signal encoding and transmission, as well as stimulation method employed to deliver the feedback information to the user of a prosthesis (Table 9.2).

To the best of our knowledge, the study [79] is the sole example of closing the loop in prosthetics using a biomimetic (scalable) skin structure coupled to a real prosthetic hand (Bebionic, Ottobock). This biologically inspired e-dermis enables the prosthesis user to perceive a continuous spectrum from innocuous to noxious touch through a neuromorphic interface that produces receptor-like spiking neural activity. A biomimetic multilayered e-dermis is placed over the fingertips of the thumb, index, ring, and little fingers: conductive and piezoresistive textile sensors are encased in rubber. A dermal layer of two piezoresistive sensing elements is separated from the epidermal layer (with one piezoresistive sensing element, only) with a 1-mm layer of silicone rubber. The epidermal layer is more sensitive and has a larger change in resistance compared with the dermal layer. The e-dermis output is used as input current to the artificial neuron model, which provides information on the loaded fingers with three levels of pressure using neuromorphic representation of touch. The neural equivalent of a tactile signal is fed back to the user through transcutaneous electrical nerve stimulation. The median and ulnar nerves, innervating relevant areas of the phantom hand, have been identified on the amputee's left residual limb and targeted for noninvasive electrical stimulation. Moreover, an autonomous pain reflex is activated on the prosthetic device, mimicking the functionality of the polysynaptic pain reflex in humans. Though sensors are embedded into a compliant multilayered structure mimicking the human skin, the drawback of this solution is still the low number of sensors, which prevents higher resolution sensing and more human-like tactile perception.

Research illustrated in [80] is the first attempt for sophisticated electrotactile feedback integrating advanced distributed sensing (e-skin including 64 piezoelectric polymer sensors) and distributed electrostimulation (32 electrodes). For sensing, piezoelectric polymer sensors (PVDF) have been used, which might be considered as biomimetic in that they have the characteristics of both FA and SA receptors in the human skin (see previous

Table 9.2 Electronic skin systems coupled with electrotactile stimulation for feedback

References	Sensing	Processing and Encoding	Transmission and Feedback	Suitable Features
[79]	**Multilayered e-dermis** placed over the thumb, index, ring, little of the **Bebionic** prosthesis (3 taxels / fingertip) Pressure sensors made of piezoresistive and conductive fabrics Sensing area ~1.5 cm²/fingertip	**Output**: • Pain detection • Discrimination of object curvature and sharpness • Detection of fingertip pressure, contact rate, number of active taxels • Feedback on loaded fingers with 3 pressure levels using neuromorphic representation • Prosthesis reflex after ~100 ms of pain detection	Sensor signal input to <u>prosthesis control</u> (controlled from PC via BT) <u>Stimulator control</u>: 3 levels of pressure encoded through stimulation pulse width and frequency	**Sensing** (piezoresistive) e-dermis to perceive continuous spectrum from innocuous to noxious touch through a neuromorphic interface that produces receptor-like spiking neural activity • Sensing pressures up to 300 kPa • High sensitivity **Feedback**: Transcutaneous electrical nerve stimulation (TENS) using Neuromorphic mode
[80]	**E-skin** with 64 taxels: piezoelectric (PVDF) *pressure* sensors. **Overall size**: rectangular patch (12.8 × 4.8 cm²) **Single Taxel radius**: 1.5 mm	**Processing**: • Calibration • Time integration • Spatial fusion **Output**: Detecting moving contact on the skin	Single voltage value per data segment (average on 4 neighbor taxels) sent to the stimulator at the sampling rate of 20 Hz (new value each 50 ms)	**Sensing** (piezoelectric): • Detect low pressures (>100 Pa) • Large bandwidth (0.1 Hz–1 kHz) • Mechanical flexibility

(Continued)

Table 9.2   Continued

References	Sensing	Processing and Encoding	Transmission and Feedback	Suitable Features
				Feedback: noninvasive, cutaneous electrotactile **System**: enables recognition of moving stimuli features (position, direction, shape)

section). This study was conducted as a proof of concept to demonstrate the feasibility: the system has been tested in eight healthy subjects who were asked to recognize the shape, trajectory, and direction of a set of dynamic movement patterns (single lines, geometrical objects, and letters) presented on the e-skin. The experiments proved that the system successfully translated the mechanical interaction into the moving electrotactile profiles, which the subjects could recognize with a good performance (shape recognition: $86 \pm 8\%$ lines, $73 \pm 13\%$ geometries, and $72 \pm 12\%$ letters). In particular, the subjects could identify the movement direction with a high confidence. Though coupled to a rigid substrate for the current study, the sensing system is flexible enough to be integrated on curved substrates as those of a prosthetic hand.

Such high-resolution noninvasive feedback interfaces might be used to explore novel scenarios for effective communication with the amputee. In the same context, though sensing is not included this time, a multimodal method to encode tactile information to possibly maximize information transmission to the amputee was presented in [81]. In that study, we described a novel noninvasive interface for multichannel electrotactile feedback (matrix of 24 electrodes) and investigated to what extent able-bodied human subjects could localize the electrotactile stimulus delivered through the matrix. Experimental results on eight able-bodied subjects showed that a novel dual-parameter modulation scheme (with interleaved frequency and intensity) substantially improved the spatial localization of the stimulus compared to both same-frequency and mechanical stimulations. This work thus contributes to coding approaches that can be used in combination with e-skins to noninvasively close the sensorimotor control loop in prosthetics.

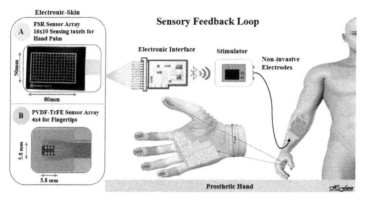

Figure 9.3 Demonstration of a complete distributed sensing-stimulation system for prosthetics applications [82]. The sensor matrix (left) is connected to acquisition electronics that samples the tactile signals and sends commands to a stimulator via a Bluetooth link. The stimulator generates electrical pulses that are delivered to the prosthesis user via a matrix electrode placed on the residual limb.

The next step will be to directly couple the e-skin to a prosthetic hand and the cutaneous multichannel electrode to the residual limb of an amputee, while preserving the described methods to encode tactile information. Figure 9.3 illustrates the overall concept, which has been demonstrated in this version in [82]. The live demo system included (i) high-resolution flexible e-skin patches (4 × 4 arrays) based on piezoelectric polymers to be integrated into the prosthesis fingertip and a commercial (16 × 10) FSR sensor array based on piezoresistive polyester as a proof of concept for palm sensors, (ii) a SoC-based electronics for signal conditioning, data acquisition and processing, and wireless stimulator control, and (iii) a fully programmable multichannel electrostimulator connected to flexible electrode matrices. The system automatically translated mechanical stimuli into stimulation profiles delivered to the participant through electrocutaneous stimulation. We demonstrated that the participant was able to identify how the e-skin surface had been touched, by receiving this information through electrostimulation only, proving therefore that a human subject was able to successfully interpret tactile sensations elicited by the proposed system.

Figure 9.4 finally integrates all e-skin systems associated with the case studies reported in this section (Tables 9.1 and 9.2), providing an overall view of distributed tactile sensing systems developed for prosthetics.

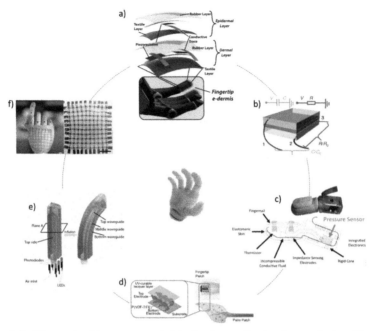

Figure 9.4 Electronic skin systems suitable for the application in prosthetics. The panels (a)–(f) depict the systems described in [56, 74–76, 79, 82].

9.3.3 Conclusive Remarks

To develop humans-in-the-loop next-generation prosthetic systems endowed with the sense of touch, advanced sensing systems that can mimic skin-like perception are needed. Beyond the state-of-the-art, human-inspired principles and current research in robotics provide valuable insights for the advancement of future general-purpose sensorimotor systems for prosthetics [61]. Novel platforms based on anthropomorphic mechanics of the artificial hand (e.g., SoftHand [83] and Hannes [84] hand) if endowed with e-skin systems might provide more functionality and utility to their users, since they would mimic the wide potentialities of the human hand for both perception and action. Managing these complex systems will require implementing sensorimotor control mechanisms that are inspired by such concepts as sensor binding, representation, dimensionality reduction, and motor and sensory synergies [61]. On the other side, a robust, intuitive, and symbiotic bidirectional interface with the human nervous system is still needed for effective brain-controlled sensorimotor operation of the prosthetic device in daily-life activities.

9.4 Electrotactile Stimulation for Sensory Feedback

The application of electrical stimulation to activate skin receptors and afferent nerves to provide somatosensory feedback is not new [22]. However, only in recent years advances in transcutaneous electrical stimulation technology, including compact stimulation hardware and multiarray electrodes, allowed the design of miniaturized systems that can provide sufficient flexibility to be considered as possibly the best channel for providing tactile feedback and closing the loop in myoelectric prostheses. Advantages of this technology compared to well-investigated vibrotactile stimulation are primarily in size. Therefore, a significantly higher number of electrodes that can serve as independent actuators can be integrated into a prosthesis socket. In addition, the stimulation parameters are decoupled, i.e., amplitude and frequency can be independently modulated and thereby serve as independent feedback channels. In vibration motors, these parameters are coupled through mechanical design (e.g., the resonance effect in a mass-spring system [85]). The primary disadvantage of electrotactile stimulation is that the elicited sensation might be inconsistent. The reported intensity, comfort, and quality can depend not only on subjective perception and individual skin properties, but also on the exact electrode location. And indeed, an important aim of the current research is to develop efficient and fast calibration procedures and electrode/skin interface that can ensure stable and intimate contact [86]. An additional disadvantage is that electrical stimuli can distort EMG signals used to control a myoelectric prosthesis [87], but there are clear indications that this can be efficiently resolved with a specific design of the multiarray stimulation electrode in combination with internal blanking algorithms. The recent findings, which will be discussed in the next section, indicate that we are getting closer to overcoming most of these inherent limitations of electrical stimulation. Therefore, a high-bandwidth tactile feedback channel that could enable real-time transfer of information from high-density e-skin sensors might be available in the near future.

9.4.1 Electrotactile Stimulation Hardware

The principle of spatial and time distribution of stimulation over multiarray electrodes was developed to address the need for more effective functional electrical stimulation systems that could allow not only easier setup and electrode placement but also greater selectivity in muscle stimulation. In functional electrical stimulation, the electrical current pulses are delivered to motor nerves to elicit their depolarization, so that they in turn activate

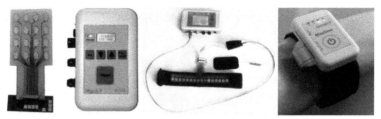

Figure 9.5   Single channel stimulation systems that integrate a demultiplexer to distribute electrical pulses to multiple electrode pads. From left to right: (a) IntFES v1 single output stage 32-channel stimulator via time multiplexing of pulses and 16-channel electrode connected to a demultiplexing unit; (b) IntFES v2 adapted for sensory stimulation connected to an array electrode via 16-channel demultiplexing unit; and (c) MaxSens system on the forearm of the subject with 16 channel of electrotactile stimulation and 8 channel of bipolar EMG recording in a compact wearable housing.

muscles to contract and produce limb movement. This approach can be used to restore movements in persons suffering from paralysis. Initially, the distributed stimulation was implemented through a switch box in which mechanical switches allowed user to manually control the output where the current would flow [88]. However, the real advantage of multiarray electrodes and distributed stimulation was achieved once a digital demultiplexing circuitry was embedded into the stimulation system. Importantly, the demultiplexer could be directly controlled by the stimulator-embedded intelligence [89]. The IntFES v1 stimulation system (Figure 9.5a), based on current controlled biphasic pulse generator, introduced the paradigm of asynchronous stimulation [90]. By distributing the stimulation pulses to several pads in a single refractory period through a demultiplexing unit, this approach allows a single output stage stimulator to effectively act like a multichannel device. For instance, IntFES v1 can be considered a 16- or 32-channel stimulator, depending on whether only one or two demultiplexing units are used. The only physical limitation to the principle of time demultiplexing of stimulation is that the number of active channels × pulse duration must be less than the stimulation period.

A clear benefit of the proposed principle demonstrated through the IntFES v1 system was that a high number of stimulation channels could be achieved with a compact wearable device, which can be controlled via a Bluetooth interface. This has provided the motivation to explore whether this technology can also be exploited in tactile feedback applications, where the goal is to activate superficial skin afferents and not the motor nerves (as in functional electrical stimulation). Therefore, a second generation of this stimulator (IntFES v2) was developed (Figure 9.5b). This system increased the number of output channels to 64 and included additional communication and control interfaces (including touch screen) to better address the requirements of tactile feedback applications. The main difference between sensory and motor stimulation is in the required current density, since the electrical charge to provoke depolarization and activity of superficial skin afferents is substantially lower compared to that needed to activate deeper efferent nerves. Therefore, the design of IntFES v2 was changed to decrease the stimulation amplitude range (0,1 to 5 mA instead of 1 to 50 mA) and the increment/decrement step to 0,1 mA, where the latter enabled a fine control of evoked sensations [91].

Several studies confirmed that a large number of stimulation channels can indeed be used to present high-bandwidth multivariable information to the user of a prosthesis [92] and that electrotactile stimulation could provide meaningful and effective feedback that can improve prosthesis control [93]. Motivated by these encouraging results, a stimulation system using time and space distribution of pulses was designed specifically for afferent stimulation (Figure 9.5c). Due to smaller current amplitudes and output voltage, the system could be further minimized. This system was conceived to enable closed-loop control of a myoelectric prosthesis, and therefore, MaxSens device integrated an 8-channel EMG amplifier together with 16 channels of distributed electrotactile stimulation. This integration also allowed the investigation and integration of various blanking algorithms that can ensure the preservation of EMG signal quality in the presence of stimulation artifacts [94].

9.4.2 Multiarray Electrodes and Electrode/Skin Interface

The systems that allow time and space distributed stimulation can be effective only when used with an adequate electrode design (Figure 9.6). For instance, if multiple channels are closer than two-point discrimination threshold [95] of intended body part, these channels will become redundant and in that sense useless for spatial coding of information. Furthermore, electrode size, shape, and the skin interface layer will affect the stimulation system requirements. For example, if the contact surface between the electrode and the skin is small, the impedance will increase and this might increase the voltage demands to the point where sufficient output current cannot be achieved. The impedance of the electrode–skin interface will not only affect hardware requirements, but the adhesive and ionic properties will also affect

Figure 9.6 Various multielectrode array configurations designed and tested for the application of electrotactile feedback, from left to right: concentric electrode 4 × 3 matrix, conductive rubber interface 2 × 8 with common anodes, and multiring electrode. This technology is very flexible and allows printing the electrodes with an arbitrary size, shape, configuration, and number of electrode pads.

stimulation comfort. All in all, the electrode design, i.e., pad size, shape, and arrangement, and the interface material need to be thoroughly investigated and adapted to the specific application of interest.

In our previous work, we tested two reference designs of the array electrode with enough pads to provide rich feedback, potentially even multiple feedback variables in parallel. The aim in both cases was to minimize the overall electrode size to allow incorporation in the socket while assuring that the distance between the pads was above the two-point discrimination threshold for the forearm (~9–10 mm [95]). First electrode design considered common anode stretching across the electrode array, while the second considered concentric electrode configuration (i.e., the anode encircling each cathode). It should be noted that in case of symmetric biphasic stimulation, the term anode is used colloquially to denote the electrode which will not be depolarized by the compensatory pulse due to a greater surface that will decrease the current density below the activation threshold. The experimental evaluation showed that both designs exhibited similar performance in stimuli localization and could be therefore effectively used for spatial coding of feedback from the myoelectric hand (Figure 9.7) [91].

9.4.3 Electrotactile Feedback From Myoelectric Prostheses

Flexibility of the distributed stimulation over multielectrode arrays allows generating versatile stimulation profiles and thereby transmitting high-bandwidth feedback information. The question is how to optimize the use of this information channel, how to encode and present the feedback variable to the user, and how to optimize the dynamic characteristics of the feedback

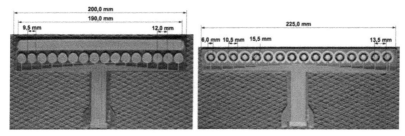

Figure 9.7 Two array electrodes designs (16 pads) for spatial information coding of feedback from a myoelectric prosthesis: common anode configuration (left), and concentric electrode configuration (right). Figure taken from [91].

so that it can be used robustly in prosthesis control. The information can be encoded by modulating the stimulation parameters (amplitude, frequency, and pulse width) and/or the active pad (the so-called spatial coding) [96]. Considering that user perception is dependent on the electrode/skin interface and location, proposed practice in the design of multivariable coding schemes is to use stimulation amplitude calibration [86]. The calibration ensures that the stimulation is not unpleasant (below the discomfort threshold) while at the same time it evokes a clear sensation (above detection threshold) for each electrode pad.

Parameter modulation and spatial coding can be combined to encode one or more independent variables from the prosthesis. The so-called mixed coding allows eliciting many different tactile sensations (N frequencies/intensities × M pads) that are still clearly discriminable. Therefore, this enables providing a high-resolution feedback when conveying a single feedback variables (e.g., grasping force). As demonstrated in [92], such feedback allows the user to modulate the prosthesis grasping force to track a reference trajectory and to grasp objects by exerting desired force levels. Furthermore, the feedback can improve the prosthesis control by facilitating short- and long-term learning [93].

However, possibly the most important advantage of multipad electrotactile feedback is that its flexibility can be used to present multivariable information simultaneously. Based on this principle, the idea of dynamic stimulation patterns was introduced [91]. The patterns were designed to convey feedback on the applied force using frequency modulation in parallel to proprioceptive feedback (wrist rotation), which is encoded through spatial modulation. Initial results from the assessment of this approach in six amputees confirmed that this type of feedback coding is indeed intuitive

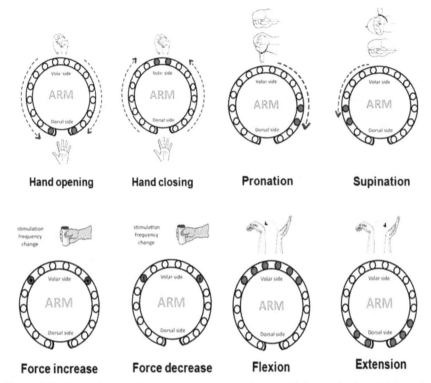

Figure 9.8   Dynamic encoding patterns to communicate multiple prosthesis variables using an array electrode that is placed around the residual limb. Hand opening and closing are indicated by moving the two electrotactile stimuli toward and away from each other. Wrist rotation is conveyed by moving the electrotactile stimulus around the forearm. Grasping force is encoded through the frequency of stimulation. The proposed patterns can be combined (superposed) to transmit several variables in parallel. Figure adapted from [91].

and allows the users to clearly differentiate feedback about four prosthesis variables (Figure 9.8).

9.5 Discussion

In the present chapter, a concept is proposed for high-bandwidth closed-loop interfacing between the user and his/her prosthetic limb (Figures 9.1 and 9.2), and the technologies that are required to implement this challenging vision are reviewed. As described in the earlier sections, there are promising developments in the field of e-skins, which are suitable for wearable applications, as well as in the design of electrocutaneous stimulation systems delivering electrical pulses through matrix electrodes with a high number of conductive pads. However, although the development of the technology is indeed essential to enable this novel approach to human–machine interfacing, there are also additional important challenges that need to be tackled before a successful application is possible. One of the critical questions is how to transmit the feedback information most effectively to the prosthesis user.

On one side, the e-skin covering the prosthetic hand with a dense network of tactile sensors will provide an abundance of tactile data. This information needs to be transmitted through the human tactile channel by delivering electrocutaneous stimulation. Considering the confined space that is available in a prosthetic socket, where in some cases the residual limb can be rather short, it is quite possible that the number of electrodes can be even substantially smaller than the number of taxels. Therefore, it is likely that some form of sensor data fusion and/or dimensionality reduction will need to be applied to the raw e-skin data. In our recent work, for example, the data from the four neighboring taxels were averaged and then mapped to a single pad within the matrix electrode [80]. The same work also demonstrates that the tactile signals might need to be conditioned and processed to extract useful aspects of tactile interaction so that those can be transmitted to the subject. In [80], the sensor signals had to be integrated to estimate the static response.

After such initial manipulations, the sensor data could be directly transmitted to the subject. In this case, the mapping between the sensor readings and electrotactile stimulation would be rather straightforward: a response of a single taxel or a group of neighboring taxels leads to activation of a pad or a group of pads in the electrode. The assumption of this approach would be that the brain could learn to extract higher-level perceptual properties from the spatially distributed stimulation delivered over the residual limb. After repeatedly grasping different objects using prosthesis and associating the elicited tactile sensations with the concomitant visual input, in one moment the user would be able to recognize the shape of the object that is grasped by focusing solely on the tactile sensations. This would hopefully include estimating not only object shape, but also other properties and aspects of tactile interaction. For example, the subject might be able to detect that the object started slipping from the grasp by registering the movement of the electrotactile stimuli from pad to pad. The previous works [97, 98] in using sensory substitution to convey visual and auditory information through the sense of touch are encouraging since they demonstrate that indeed complex information can be transmitted haptically.

Another approach would be to first process the raw sensor data to extract meaningful high-level information, which is then communicated to the subject through a matrix electrode. To this aim, material physics and machine learning [99, 100] could be applied within the embedded skin electronics to detect the type of mechanical interaction (e.g., rolling or sliding), estimate object properties, compute force vectors, and/or detect slippage. The electrotactile interface would then convey the computed outcomes instead of the raw sensor data. For instance, instead of the subject detecting slippage by noticing the movement of the electrotactile stimulus, the slippage in this approach would be indicated by activating a specific electrode pad dedicated to conveying the slippage event. The flexibility of the matrix interface can be exploited to transmit different high-level feedback variables in sequence or even simultaneously. For example, some pads can be assigned to the grasping force while the others can be used to map joint angles, hence providing artificial extero and proprioception.

The aforementioned flexibility of the interface is an advantage. However, it is also a challenge since there are many possible ways to encode feedback variables. The pads can be activated in different spatial combinations while modulating the frequency and/or intensity, where the latter can be adjusted individually for each pad. Identifying the most promising combination of encoding schemes to transmit a high volume of data (taxel responses and/or multiple feedback variables) is still an open research question.

The main motivation for implementing a high-density sensing and stimulation interface is that it would improve the utility of a prosthetic system as well as user experience. A sophisticated feedback on the hand–object interaction would allow the user to command the prosthesis to achieve more stable and effective grasp. At the same time, this would open up a pathway to address additional scenarios that have not been considered before in prosthetics, such as, allowing the prosthesis user to feel passive, social, and affective touch [101]. For example, the feedback would indicate when someone else touches the prosthesis, and the amputee could perceive the sensation of gentle caressing. These types of interactions are likely to substantially facilitate the embodiment of the prosthesis.

One aspect that has not been addressed in this chapter is that the information measured by an e-skin can be also used to implement a local feedback look, within the device itself. The artificial controller could exploit the tactile data to control the prosthesis automatically. When slippage is detected, for instance, the grasping force could be increased mimicking the action of a biological reflex. This action is already implemented in some commercial systems [102] using a single discrete sensor. The e-skin would allow detecting the slippage event in more detail possibly triggering different recovery strategies (e.g., activating different fingers to prevent the slip in a particular direction). Furthermore, the information on the pattern of contacts while the prosthesis is closing around the object, provided by the e-skin, could be used to move individual fingers to achieve force closure and overall stable grasp (tactile-servoing) [103].

In this chapter, we have considered the application of advanced sensing and stimulation technologies in the realm of upper limb prostheses. Nevertheless, these methods are equally relevant for lower limb prostheses. In this case, the matrix electrode would be wrapped around the shank (transtibial amputation) or thigh (transhumeral amputee) while the e-skin would be covering the prosthetic foot. The spatially distributed stimulation could be used, for instance, to convey the trajectory of the center of pressure as it moves along the sole of the foot during the stance phase of gait. This and other continuous (e.g., magnitude of grasping force) or discrete (e.g., contact events) information could improve the stability and quality of walking in lower limb amputees.

Overall, it seems that the technological components are already available for the implementation of the proposed conceptual solution, but the critical effort still needs to be invested to implement the full high-density closed-loop solution operating online. This solution then needs to be tested psychometrically as well as in functional tasks to assess the impact of such feedback on the interaction between the user and his/her bionic limb.

9.6 Conclusions

Motivated by the developments of the artificial skin sensors and compact multichannel stimulation devices, this chapter proposes a conceptual solution for a high-bandwidth feedback interface between the user and his/her bionic limb. The state-of-the-art technologies as well as initial efforts to close the control loop are reviewed, and the remaining challenges such as variable selection and encoding of information are discussed. The proposed concept can potentially revolutionize the interaction between the user and his/her prosthesis by providing spatially distributed tactile stimulation, mimicking thereby the natural feedback provided by the sound hands. This can facilitate prosthesis control and improve user experience, and enable touch modalities that have not been considered in prosthetics before (e.g., the implementation of passive, social, and affective touch).

References

[1] C. MacKenzie and T. Iberall, *The Grasping Hand*. Amsterdam: Elsevier B.V, 2010.

[2] J. R. Flanagan, S. King, D. M. Wolpert, and R. S. Johansson, "Sensorimotor prediction and memory in object manipulation," *Can. J. Exp. Psychol. Can. Psychol. Expérimentale*, vol. 55, no. 2, pp. 87–95, 2001.

[3] B. Gesslbauer, L. A. Hruby, A. D. Roche, D. Farina, R. Blumer, and O. C. Aszmann, "Axonal components of nerves innervating the human arm," *Ann. Neurol.*, vol. 82, no. 3, pp. 396–408, Sep. 2017.

[4] R. Shadmehr and J. W. Krakauer, "A computational neuroanatomy for motor control," *Exp. Brain Res.*, vol. 185, no. 3, pp. 359–81, Mar. 2008.

[5] M. S. Johannes et al.,"The modular prosthetic limb," in *Wearable Robotics*, Academic Press, Elsevier, pp. 393–444, 2020.

[6] L. Resnik, S. L. Klinger, and K. Etter, "The DEKA Arm: Its features, functionality, and evolution during the Veterans Affairs Study to optimize the DEKA Arm," *Prosthet. Orthot. Int.*, pp. 0309364613506913-, Oct. 2013.

[7] J. T. Belter, J. L. Segil, A. M. Dollar, and R. F. Weir, "Mechanical design and performance specifications of anthropomorphic prosthetic hands: A review," *J. Rehabil. Res. Dev.*, vol. 50, no. 5, pp. 599–618, Jan. 2013.

[8] "Coapt//Home Page." [Online]. Available from http://coaptengineering.com/. [accessed on 05-Apr-2015].

[9] "MyoPlus Otto Bock." [Online]. Available from https://www.ottobockus.com/prosthetics/upper-limb-prosthetics/solution-overview/myo-plus/myo-plus.html. [accessed on 29-May-2020].

[10] F. Cordella et al., "Literature Review on Needs of Upper Limb Prosthesis Users," Front. Neurosci., vol. 10, May 2016.

[11] J. S. Schofield, K. R. Evans, J. P. Carey, and J. S. Hebert, "Applications of sensory feedback in motorized upper extremity prosthesis: A review," *Expert Rev. Med. Devices*, vol. 11, no. 5, pp. 499–511, Sep. 2014.

[12] C. Antfolk, M. D'Alonzo, B. Rosén, G. Lundborg, F. Sebelius, and C. Cipriani, "Sensory feedback in upper limb prosthetics," *Expert Rev. Med. Devices*, vol. 10, no. 1, pp. 45–54, Jan. 2013.

[13] P. Svensson, U. Wijk, A. Björkman, and C. Antfolk, "A review of invasive and non-invasive sensory feedback in upper limb prostheses," *Expert Rev. Med. Devices*, vol. 14, no. 6, pp. 439–447, Jun. 2017.

[14] "Vincent Evolution 2 Hand," 2019. [Online]. Available from http://vi ncentsystems.de/en/prosthetics/vincent-evolution-2/. [accessed on 09-Oct-2014].

[15] "Ability hand." [Online]. Available from https://www.psyonic.co/abili tyhand. [accessed on 29-May-2020].

[16] D. W. Tan, M. A. Schiefer, M. W. Keith, J. R. Anderson, J. Tyler, and D. J. Tyler, "A neural interface provides long-term stable natural touch perception," *Sci. Transl. Med.*, vol. 6, no. 257, pp. 257ra138–257ra138, Oct. 2014.

[17] E. D'Anna et al., "A closed-loop hand prosthesis with simultaneous intraneural tactile and position feedback," *Sci. Robot.*, vol. 4, no. 27, p. eaau8892, Feb. 2019.

[18] J. A. George et al., "Biomimetic sensory feedback through periph-eral nerve stimulation improves dexterous use of a bionic hand," *Sci. Robot.*, vol. 4, no. 32, p. eaax2352, Jul. 2019.

[19] K. R. Schoepp, M. R. Dawson, J. S. Schofield, J. P. Carey, and J. S. Hebert, "Design and integration of an inexpensive wearable mechan-otactile feedback system for myoelectric prostheses," *IEEE J. Transl. Eng. Heal. Med.*, vol. 6, pp. 1–11, 2018.

[20] S. Casini, M. Morvidoni, M. Bianchi, M. Catalano, G. Grioli, and A. Bicchi, "Design and realization of the CUFF – clenching upper-limb force feedback wearable device for distributed mechano-tactile stimulation of normal and tangential skin forces," in *2015 IEEE/RSJ International Conference on Intelligent Robots and Systems (IROS)*, Hamburg, Germany, October 2015, pp. 1186–1193.

[21] K. A. Kaczmarek, J. G. Webster, P. Bach-y-Rita, and W. J. Tomp-kins, "Electrotactile and vibrotactile displays for sensory substitution systems," *IEEE Trans. Biomed. Eng.*, vol. 38, no. 1. Ieee, pp. 1–16, 1991.

[22] A. Y. Szeto and F. A. Saunders, "Electrocutaneous stimulation for sensory communication in rehabilitation engineering," *IEEE Trans. Biomed. Eng.*, vol. 29, no. 4, pp. 300–8, Apr. 1982.

[23] J. W. Sensinger and S. Dosen, "A Review of Sensory Feedback in Upper-Limb Prostheses From the Perspective of Human Motor Control," *Front. Neurosci.*, vol. 14, p. 345, Jun. 2020.

[24] J. S. Schofield, C. E. Shell, D. T. Beckler, Z. C. Thumser, and P. D. Marasco, "Long-term home-use of sensory-motor-integrated bidi-rectional bionic prosthetic arms promotes functional, perceptual, and cognitive Changes," *Front. Neurosci.*, vol. 14, p. 120, Feb. 2020.

[25] M. Markovic et al., "The clinical relevance of advanced artificial feedback in the control of a multi-functional myoelectric prosthesis," *J. Neuroeng. Rehabil.*, vol. 15, no. 1, p. 28, Mar. 2018.

[26] A. W. Shehata, M. Rehani, Z. E. Jassat, and J. S. Hebert, "Mechanotactile sensory feedback improves embodiment of a prosthetic hand during active use," *Front. Neurosci.*, vol. 14, p. 263, Mar. 2020.

[27] R. Dahiya, D. Akinwande, and J. S. Chang, "October 2019: Flexible electronic skin: From humanoids to humans [scanning the issue]," *Proc. IEEE*, vol. 107, no. 10, pp. 2011–2015, 2019.

[28] E. Gardner and K. Johnson, "The somatosensory system: receptors and central pathways," in *Principles of Neural Science*, 5th ed., E. Kandel, J. Schwartz, T. Jessell, S. Siegelbaum, and A. Hudspeth, Eds. New York, NY, USA: McGraw-Hill, pp. 475–495, 2013.

[29] R. S. Johansson and J. R. Flanagan, "Coding and use of tactile signals from the fingertips in object manipulation tasks," *Nat. Rev. Neurosci.*, vol. 10, no. 5, pp. 345–359, 2009.

[30] A. Chortos, J. Liu, and Z. Bao, "Pursuing prosthetic electronic skin," *Nat. Mater.*, vol. 15, no. 9, p. 937–950, Sep. 2016.

[31] R. Dahiya et al., "Large-area soft e-skin: The challenges beyond sensor designs," *Proc. IEEE*, vol. 107, no. 10, pp. 2016–2033, Oct. 2019.

[32] P. Maiolino, M. Maggiali, G. Cannata, G. Metta, and L. Natale, "A flexible and robust large scale capacitive tactile system for robots," *IEEE Sens. J.*, vol. 13, no. 10, pp. 3910–3917, 2013.

[33] L. Seminara, M. Capurro, P. Cirillo, G. Cannata, and M. Valle, "Electromechanical characterization of piezoelectric PVDF polymer films for tactile sensors in robotics applications," *Sensor. Actuat., A-Phys.*, vol. 169, no. 1, pp. 49-58, 2011.

[34] X. Hen et al., "Self-powered flexible pressure sensors with vertically well-aligned piezoelectric nanowire arrays for monitoring vital signs," *J. Mater. Chem. C*, vol. 3, no. 45, pp. 11806–11814, 2015.

[35] M. Iskarous and N. V. Thakor, "E-Skins: Biomimetic sensing and encoding for upper limb prostheses," in *Proceedings of the IEEE*, vol. 107, no. 10, pp. 2052–2064, 2019.

[36] A. Moscatelli et al., "Touch as an auxiliary proprioceptive cue for movement control," *Sci. Adv.*, vol. 5, no. 6, p. eaaw3121, Jun. 2019.

[37] S. J. Lederman and R. L. Klatzky, "Haptic perception: A tutorial," *Atten. Percept. Psychophys.*, vol. 71, no. 7, pp. 1439–1459, Oct. 2009.

[38] I. Birznieks, V. G. Macefield, G. Westling, and R. S. Johansson, "Slowly adapting mechanoreceptors in the borders of the human

fingernail encode fingertip forces," *J. Neurosci.*, vol. 29, no. 29, pp. 9370–9379, 2009.

[39] L. Seminara, M. Valle, and M. Capurro, "Bending response of PVDF piezoelectric sensors," *SENSORS, 2012 IEEE*, Taipei, 2012, pp. 1–4. doi: 10.1109/ICSENS.2012.6411405.

[40] P. Jenmalm, I. Birznieks, A. W. Goodwin, and R. S. Johansson, "Influence of object shape on responses of human tactile afferents under conditions characteristic of manipulation," *Eur. J. Neurosci.*, vol. 18, no. 1, pp. 164–176, 2003.

[41] M. J. Adams, S. A. Johnson, P. Lefèvre, V. Lévesque, V. Hayward, T. André, et al., "Finger pad friction and its role in grip and touch," *J. R. Soc. Interface*, vol. 10, no. 80, pp. 1–19, 2013.

[42] J. Scheibert, S. Leurent, A. Prevost, G. Debrégeas, "The role of fingerprints in the coding of tactile information probed with a biomimetic sensor," *Science*, vol. 323, no. 5920, pp. 1503–1506, 2009.

[43] M. L. Hammock, A. Chortos, B. C.-K. Tee, J. B.-H. Tok, and Z. Bao, "25th Anniversary Article: The evolution of electronic skin (E-skin): A brief history, design considerations, and recent progress," *Adv. Mater.*, vol. 25, no. 42, pp. 5997–6038, Nov. 2013.

[44] T. Someya et al., "Conformable, flexible, large-area networks of pressure and thermal sensors with organic transistor active matrixes," *Proc. Natl. Acad. Sci.*, vol. 102, no. 35, pp. 12321–12325, 2005.

[45] J. Kang, J. B. Tok, and Z. Bao, "Self-healing soft electronics," *Nat. Electron.*, vol. 2, pp. 144–150, 2019.

[46] J. A. Rogers, T. Someya, and Y. Huang, "Materials and mechanics for stretchable electronics," *Science.*, vol. 327, no. 5973, pp. 1603–1607, 2010.

[47] D.-H. Kim, J.-H. Ahn, W. M. Choi, H.-S. Kim, T.-H. Kim, J. Song, et al., "Stretchable and foldable silicon integrated circuits," *Science.*, vol. 320, no. 5875, pp. 507–511, 2008.

[48] S. Khan, L. Lorenzelli, and R. S. Dahiya, "Technologies for printing sensors and electronics over large flexible substrates: A review," *IEEE Sens. J.*, vol. 15, no. 6, pp. 3164–3185, 2015.

[49] H. Chen, Y. Cao, J. Zhang, and C. Zhou, "Large-scale complementary macroelectronics using hybrid integration of carbon nanotubes and IGZO thin-film transistors," *Nat. Commun.*, vol. 5, no. 1, pp. 1–12, 2014.

[50] J.-J. Cabibihan, D. Joshi, Y. M. Srinivasa, M. A. Chan, and A. Muruganantham, "Illusory sense of human touch from a warm and soft

artificial hand," *IEEE Trans. Neural Syst. Rehabil. Eng.*, vol. 23, no. 3, pp. 517–527, 2014.

[51] K.-I. Jang et al., "Soft network composite materials with deterministic and bio-inspired designs," *Nat. Commun.*, vol. 6, no. 1, pp. 1–11, 2015.

[52] F. Mancini et al., "Whole-body mapping of spatial acuity for pain and touch," *Ann. Neurol.*, vol. 75, no. 6, pp. 917–924, 2014.

[53] Q. Sun, W. Seung, B. J. Kim, S. Seo, S.-W. Kim, and J. H. Cho, "Active matrix electronic skin strain sensor based on piezopotential-powered graphene transistors," *Adv. Mater.*, vol. 27, no. 22, pp. 3411–3417, 2015.

[54] K. O. Johnson and S. S. Hsiao, "Evaluation of the relative roles of slowly and rapidly adapting afferent fibers in roughness perception," *Can. J. Physiol. Pharmacol.*, vol. 72, no. 5, pp. 488–497, 1994.

[55] S. J. Bensmaïa and M. Hollins, "The vibrations of texture," *Somatosens. Mot. Res.*, vol. 20, no. 1, pp. 33–43, 2003.

[56] J. A. Fishel and G. E. Loeb, "Bayesian exploration for intelligent identification of textures," *Front. Neurorobot.*, vol. 6, p. 4, Jan. 2012.

[57] G. Loeb and J. Fishel, "The role of fingerprints in vibrotactile discrimination," *White Pap. DoD Phys. Biol.*, 2009.

[58] S. Salehi, J.-J. Cabibihan, and S. S. Ge, "Artificial skin ridges enhance local tactile shape discrimination," *Sensors*, vol. 11, no. 9, pp. 8626–8642, 2011.

[59] B. P. Delhaye, E. W. Schluter, and S. J. Bensmaia, "Robopsychophysics: Extracting behaviorally relevant features from the output of sensors on a prosthetic finger," *IEEE Trans. Haptics*, vol. 9, no. 4, pp. 499–507, 2016.

[60] R. Johansson and I. Birznieks, "First spikes in ensembles of human tactile afferents code complex spatial fingertip events," *Nat. Neurosci.*, vol. 7, pp. 170–177, 2004.

[61] L. Seminara, P. Gastaldo, S. J. Watt, K. F. Valyear, F. Zuher, and F. Mastrogiovanni, "Active haptic perception in robots: A review," *Front. Neurorobot.*, vol. 13, p. 53, Jul. 2019.

[62] A. Ibrahim and M. Valle, "Real-time embedded machine learning for tensorial tactile data processing," *IEEE Trans. Circuits Syst. I Regul. Pap.*, vol. 65, no. 11, pp. 3897–3906, 2018.

[63] A. Ibrahim, M. Osta, M. Alameh, M. Saleh, H. Chible, and M. Valle, "Approximate computing methods for embedded machine learning," in *25th IEEE International Conference on Electronics, Circuits and Systems (ICECS)*, 2018, pp. 845–848.

[64] M. Osta et al., "An energy efficient system for touch modality classification in electronic skin applications," *2019 IEEE International Symposium on Circuits and Systems (ISCAS)*, Sapporo, Japan, pp. 1–4, 2019. doi: 10.1109/ISCAS.2019.8702113.

[65] J. G. Webster et al., "Neuromorphic systems," in *Wiley Encyclopedia of Electrical and Electronics Engineering*, John Wiley & Sons, Inc., New York, 2016, pp. 1–22.

[66] S. Caviglia, M. Valle, and C. Bartolozzi, "Asynchronous, event-driven readout of POSFET devices for tactile sensing," in *2014 IEEE International Symposium on Circuits and Systems (ISCAS)*, Melbourne VIC, 2014, pp. 2648–2651, doi: 10.1109/ISCAS.2014.6865717.

[67] F. Bergner, E. Dean-Leon, and G. Cheng, "Design and realization of an efficient large-area event-driven E-skin," *Sensors*, vol. 20, no. 7, p. 1965, Mar. 2020.

[68] B. C.-K. Tee et al., "A skin-inspired organic digital mechanoreceptor," *Science.*, vol. 350, no. 6258, pp. 313–316, 2015.

[69] W. Wei Lee et al., "A neuro-inspired artificial peripheral nervous system for scalable electronic skins," *Sci. Robot.*, vol. 4, no. 32, eaax2198, 2019.

[70] C. García Núñez, L. Manjakkal, and R. Dahiya, "Energy autonomous electronic skin," *npj Flex. Electron.*, vol. 3, no. 1, pp. 1–24, Dec. 2019.

[71] Q. Sun et al., "Transparent, low-power pressure sensor matrix based on coplanar-gate graphene transistors," *Adv. Mater.*, vol. 26, no. 27, pp. 4735–4740, 2014.

[72] M. Shi et al., "Self-powered analogue smart skin," *ACS Nano*, vol. 10, no. 4, pp. 4083–4091, 2016.

[73] PH-CODING: Towards a revolution in the understanding of human haptics, European Union's Horizon 2020 FET Open programme, grant agreement. No. 829186, https://www.ph-coding.eu/.

[74] Z. Lei and P. Wu, "A supramolecular biomimetic skin combining a wide spectrum of mechanical properties and multiple sensory capabilities," *Nat. Commun.*, vol. 9, p. 1134, 2018.

[75] Q. Hua et al., "Skin-inspired highly stretchable and conformable matrix networks for multifunctional sensing," *Nat. Commun.*, vol. 9, p. 244, 2018.

[76] H. Zhao, K. O'Brien, S. Li, and R. F. Shepherd, "Optoelectronically innervated soft prosthetic hand via stretchable optical waveguides," *Sci. Robot.*, vol. 1, no. 1, p. eaai7529, 2016.

[77] Z. Su, J. A. Fishel, T. Yamamoto, and G. Loeb, "Use of tactile feedback to control exploratory movements to characterize object compliance," *Front. Neurorobot.*, vol. 6, p. 7, 2012.

[78] B. Matulevich, G. E. Loeb, and J. A. Fishel, "Utility of contact detection reflexes in prosthetic hand control," in *2013 IEEE/RSJ International Conference on Intelligent Robots and Systems (IROS)*, Tokyo, 2013, pp. 4741–4746. doi: 10.1109/IROS.2013.6697039.

[79] L. E. Osborn et al., "Prosthesis with neuromorphic multilayered e-dermis perceives touch and pain," *Sci. Robot.*, vol. 3, no. 19, p. eaat3818, Jun. 2018.

[80] M. Franceschi, L. Seminara, S. Dosen, M. Strbac, M. Valle, and D. Farina, "A system for electrotactile feedback using electronic skin and flexible matrix electrodes: Experimental evaluation," *IEEE Trans. Haptics*, vol. 10, no. 2, pp. 162–172, Apr. 2017.

[81] L. Seminara et al., "Dual-parameter modulation improves stimulus localization in multichannel electrotactile stimulation," *IEEE Trans. Haptics*, vol. 13, no. 2, pp. 393–403, 2020.

[82] M. Alameh et al., "Live demonstration: System based on electronic skin and cutaneous electrostimulation for sensory feedback in prosthetics," in *2018 IEEE Biomedical Circuits and Systems Conference (BioCAS)*, Cleveland, OH, 2018, p. 1. doi: 10.1109/BIOCAS.2018.8584710.

[83] S. B. Godfrey et al., "SoftHand at the CYBATHLON: A user's experience," *J. Neuroeng. Rehabil.*, vol. 14, no. 1, p. 124, 2017.

[84] https://designawards.core77.com/health-wellness/87102/Hannes-prosthetic-hand

[85] M. Azadi and L. A. Jones, "Vibrotactile actuators: Effect of load and body site on performance," in *2014 IEEE Haptics Symposium (HAPTICS)*, Houston, TX, 2014, pp. 351–356. doi: 10.1109/HAPTICS.2014.6775480.

[86] M. Isaković, J. Malešević, T. Keller, M. Kostić, and M. Štrbac, "Optimization of semiautomated calibration algorithm of multichannel electrotactile feedback for myoelectric hand prosthesis," *Appl. Bionics Biomech.*, vol. 2019, pp. 1–9, Mar. 2019.

[87] C. Hartmann, S. Dosen, S. Amsuess, and D. Farina, "Closed-loop control of myoelectric prostheses with electrotactile feedback: Influence of stimulation artifact and blanking," *IEEE Trans. Neural Syst. Rehabil. Eng.*, vol. 23, no. 5, pp. 807–816, Sep. 2015.

[88] G. Bijelic, A. Popovic-Bijelic, N. Jorgovanovic, D. Bojanic, and D. Popovic, "E Actitrode: The new selective stimulation interface for functional movements in hemiplegics patients," *Serbian J. Electr. Eng.*, vol. 1, no. 3, pp. 21–28, 2004.

[89] R. Velik, N. Malešević, L. Z. Popović, U. Hoffmann, and T. Keller, "INTFES: A multi-pad electrode system for selective transcutaneous electrical muscle stimulation," in Proc. *IFESS*, São Paulo, Brasil 8–11 September 2011.

[90] N. M. Malešević et al., "A multi-pad electrode based functional electrical stimulation system for restoration of grasp," *J. Neuroeng. Rehabil.*, vol. 9, no. 1, p. 66, Jan. 2012.

[91] M. Štrbac et al., "Integrated and flexible multichannel interface for electrotactile stimulation," *J. Neural Eng.*, vol. 13, no. 4, p. 046014, Aug. 2016.

[92] S. Dosen et al., "Multichannel electrotactile feedback with spatial and mixed coding for closed-loop control of grasping force in hand prostheses," *IEEE Trans. Neural Syst. Rehabil. Eng.*, vol. 25, no. 3, pp. 183–195, Mar. 2017.

[93] M. Strbac et al., "Short- and long-term learning of feedforward control of a myoelectric prosthesis with sensory feedback by amputees," *IEEE Trans. Neural Syst. Rehabil. Eng.*, vol. 25, no. 11, pp. 2133–2145, Nov. 2017.

[94] M. Strbac et al., "Compact electrotactile stimulation and EMG acquisition system with embedded blanking of stimulation artifacts," in *2019 26th IEEE International Conference on Electronics, Circuits and Systems (ICECS)*, Genoa, Italy, 2019, pp. 863–866. doi: 10.1109/ICECS46596.2019.8965115.

[95] M. Solomonow, J. Lyman, and A. Freedy, "Electrotactile two-point discrimination as a function of frequency, body site, laterality, and stimulation codes," *Ann. Biomed. Eng.*, vol. 5, no. 1, pp. 47–60, Mar. 1977.

[96] M. Perovic et al., "Electrical stimulation of the forearm: A method for transmitting sensory signals from the artificial hand to the brain," *J. Autom. Control*, vol. 21, no. 1, pp. 13–18, 2013.

[97] P. Bach-y-Rita and S. W. Kercel, "Sensory substitution and the human–machine interface," *Trends Cogn. Sci.*, vol. 7, no. 12, pp. 541–546, Dec. 2003.

[98] K. A. Kaczmarek, J. G. Webster, P. Bach-y-Rita, and W. J. Tompkins, "Electrotactile and vibrotactile displays for sensory substitution

systems," *IEEE Trans. Biomed. Eng.*, vol. 38, no. 1, pp. 1–16, Jan. 1991.

[99] L. Seminara, M. Capurro, and M. Valle, "Tactile data processing method for the reconstruction of contact force distributions," *Mechatronics*, vol. 27, pp. 28–37, Apr. 2015.

[100] P. Gastaldo, L. Pinna, L. Seminara, M. Valle, and R. Zunino, "A Tensor-based pattern-recognition framework for the interpretation of touch modality in artificial skin systems," *IEEE Sens. J.*, vol. 14, no. 7, pp. 2216–2225, Jul. 2014.

[101] P. Beckerle et al., "Feel-good robotics: Requirements on touch for embodiment in assistive robotics," *Front. Neurorobot.*, vol. 12, Dec. 2018. doi: 10.3389/fnbot.2018.00084

[102] "Sensor Hand." [Online] Accessed on 03.11.2020. Available from http://www.ottobock.com.au/prosthetics/upper-limb/solution-overview/myoeletric-devices-speedhands/.

[103] K. Z. Zhuang et al., "Shared human–robot proportional control of a dexterous myoelectric prosthesis," *Nat. Mach. Intell.*, vol. 1, no. 9, pp. 400–411, Sep. 2019.

Index

About the Editors

Ali Ibrahim received his M.S. degree in industrial control from the Doctoral School of Sciences and Technologies, Lebanese University, in 2009, and the Ph.D. degree in electronic and computer engineering and robotics and telecommunications from the University of Genoa-Italy in 2016. From 2009 to 2013, he was a project designer in electronics in Beirut. He is currently an assistant professor at the Lebanese International University, and an associate researcher with the Department of Electric, Electronic, Telecommunication Engineering and Naval Architecture, University of Genoa. His research interests involve embedded electronic systems, field programmable gate array (FPGA) implementation, digital data processing, interface electronics for electronic-skin systems, embedded machine learning, approximate computing, and techniques and methods for energy efficient embedded computing.

Maurizio Valle (MV) received his M.S. degree in electronic engineering in 1985 and the Ph.D. degree in electronics and computer science in 1990 from the University of Genova, Italy. In 1992 he joined the University of Genova, first as an assistant and in 2007 as an associate professor. Since December 2019, MV is full professor at the Department of Electrical, Electronic and Telecommunications Engineering and Naval Architecture, University of Genova, and leads the Connected Objects, Smart Materials, Integrated Circuits – COSMIC laboratory. MV has been and is in charge of many research contracts and projects funded at local, national, and European levels and by Italian and foreign companies. Professor Valle is a coauthor of more than 200 papers on international scientific journals and conference proceedings and of the book titled, *Robotic Tactile Sensing Technologies and System*, Springer Science+Business Media, Dordrecht, pp. 1–248, 2013 (ISBN: 978-94007-0578-4). His research interests include electronic and microelectronic systems, material integrated sensing systems, tactile sensors, and electronic-skin systems. He is an IEEE senior member and also a member of the IEEE CAS Society.

For Product Safety Concerns and Information please contact our EU
representative GPSR@taylorandfrancis.com
Taylor & Francis Verlag GmbH, Kaufingerstraße 24, 80331 München, Germany